Calving the Cow
and Care of the Calf

Calving the Cow
and Care of the Calf

Eddie Straiton

The Crowood Press

First published in 1965 by
Farming Press Books, Ipswich, as
The TV Vet Book for Stock Farmers No. 2

Fourth edition 1994, retitled as
Calving the Cow and Care of the Calf

This edition published in 2002 by
The Crowood Press Ltd
Ramsbury, Marlborough
Wiltshire SN8 2HR

British Library Cataloguing-in-Publication Data
A catalogue record for this book is available from the British Library.

ISBN 1 86126 479 8

Typeset by Typestylers Ltd, Ipswich

Printed and bound in Malaysia by Times Offset (M) Sdn. Bhd.

Contents

CARE OF THE CALF

Foreword

Many years ago I was working on a farm in New Zealand. One of our house cows developed what I now know to be milk fever, but to us the illness was a mystery. Being fresh from England, I said to the boss, ''Shall I get the vet?'', to which he replied, ''Don't talk nonsense boy, a vet costs money, cut its ruddy throat, we'll feed it to the dogs.''

Times have changed now and a vet's bill is usually a better investment than the knackerman's cheque. But like many farmers, I like to know what is going on, even if I prefer to let the vet do the dirty work for me.

In this book Eddie Straiton shows in clear photographs and simple text all the problems, and their solutions, to be found when calving cows and rearing calves.

I learned my animal midwifery the hard way by practical experience, and it was particularly hard on the animals, I fear. There is no substitute for practical experience, but with this book, any young farmer and — dare I say it — veterinary surgeon, too, will be greatly helped in understanding symptoms and effecting cures. I wish the book well.

John Cherrington
Tangley, near Andover, Hampshire

Preface

This is 'it' — a concise catalogue of all the invaluable commonsense practical facts concerning calving and calves which I have learned the hard way throughout my career in veterinary practice. Nothing is written from theory — only from personal observation and practical experience, and I'm sure everyone will agree that there is no substitute for experience.

Because of the specific text, the illustrations are vitally important and I have made full use of them.

I have used the word 'cow' as a collective term to mean cow or heifer. Also I use 'stockmen' and refer to veterinary surgeons as 'he' to cover all vets and stock people, male or female.

The intelligent stockman and farmer should find many useful hints but the student, by taking full note of everything portrayed, will save himself many hours of heartache and frustration. When I was a student, I would have given anything for even a photograph of a prolapsed uterus. To be provided with a comprehensive galaxy of vital practical illustrations would have been little short of miraculous.

I am certain that all farmers, stockmen, students and younger veterinary surgeons will find this volume of inestimable value.

Two other important points — nearly one third of all losses in cattle are associated with calving and calf conditions, and practically all the material contained in this volume is applicable to cattle and calves in every part of the world.

Again I would like to pay tribute to my photographers, Mr George Pringle, who is responsible for the black and white pictures, and to Tony Boydon, who produced the superb colour plates.

I should also like to thank all the farmers and their staff who cooperated in taking the pictures.

Eddie Straiton
1994

Calving the Cow
and Care of the Calf

Anatomy of the Cow

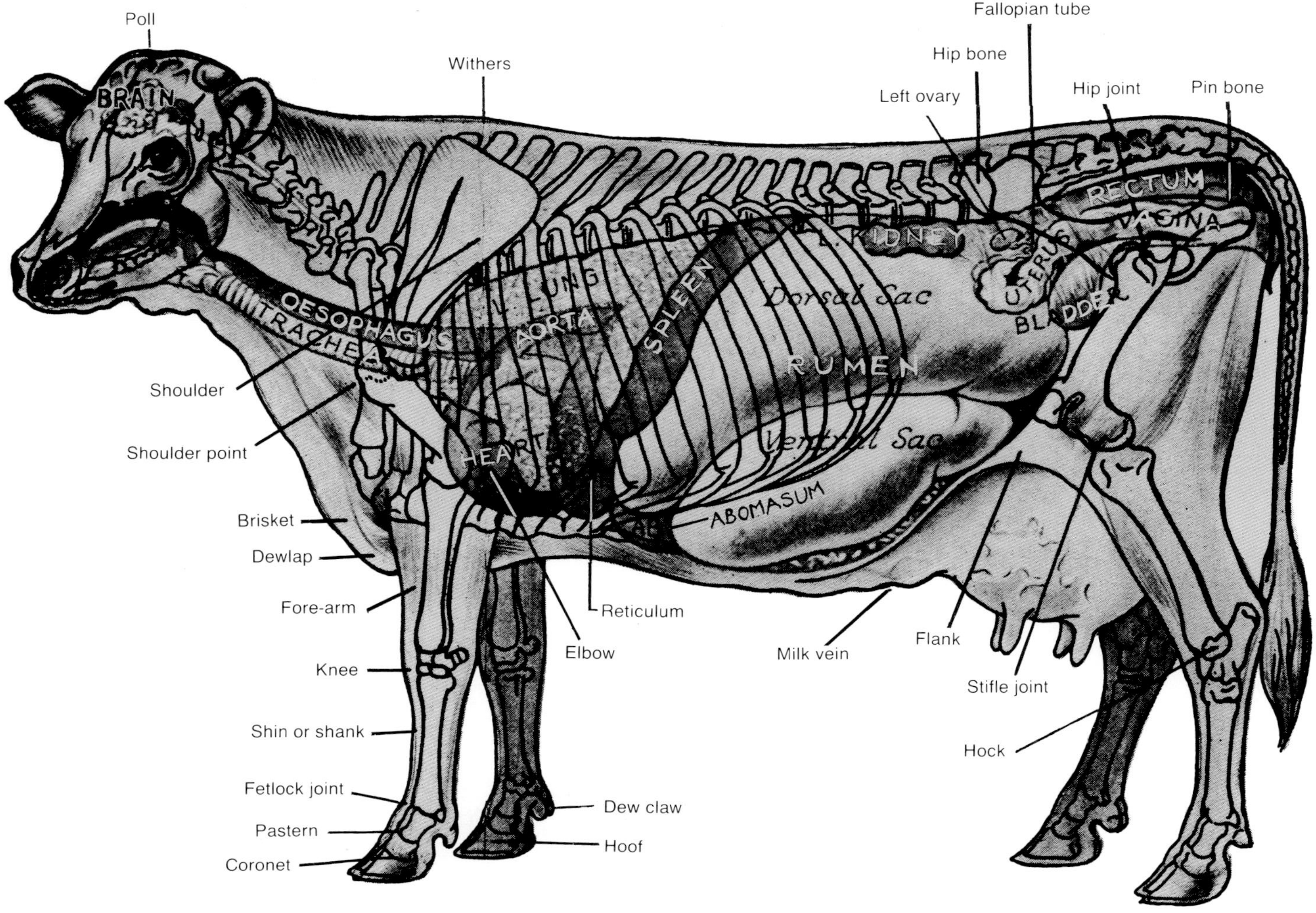

Poll

BRAIN

Withers

Fallopian tube

Hip bone

Left ovary

Hip joint

Pin bone

RECTUM

VAGINA

UTERUS

BLADDER

KIDNEY

Dorsal Sac

RUMEN

Ventral Sac

ABOMASUM

SPLEEN

LUNG

AORTA

HEART

OESOPHAGUS

TRACHEA

Shoulder

Shoulder point

Brisket

Dewlap

Fore-arm

Knee

Shin or shank

Fetlock joint

Pastern

Coronet

Reticulum

Elbow

Dew claw

Hoof

Milk vein

Flank

Stifle joint

Hock

Calving the Cow

1
Natural Birth

All my professional life I have been campaigning in both the veterinary and farming worlds for a more rational approach to the calving of heifers and cows.

Frankly, I am astonished that the strong-arm tactics of pulling the calf like a cork from a bottle ever came into being in an enlightened country. Reflect for a moment on the breeding conditions of animals in the wild and you will realise just how stupid such methods are.

Of course nature is not completely infallible, and there are times when assistance with calving is necessary, but such occasions are few. I would say without doubt that, provided the foetus is straight — and I repeat this, only when the foetus is straight — the vast majority of animals will give birth naturally if they are given a reasonable chance to do so.

It is vitally important, therefore, before ever attempting to assist at calving, to have a thorough knowledge of simple natural birth or, in other words, to really understand the job.

There are three stages in natural birth: the first or preliminary stage, the second stage, and the third or final stage.

When the cow first starts calving, and throughout the whole of the **preliminary stage**, she shows signs of intermittent uneasiness and slight pain. This is despite the fact that she may eat, drink and behave perfectly normally in every way. In fact, throughout the whole of this preliminary phase the animal is bright and relaxed and fully aware of everything going on around her.

The same, of course, holds good in humans — all through the first stage in human labour the patient will eat, drink and chat normally to people. Just one difference — where a woman may want to sit or lie back, the heifer or cow

1

stays on her feet all through the first stage.

The first sign is usually an angry swish of the tail and a restless forward movement. She will move forward in a clockwise or anticlockwise semi-circle, very often alternating one direction with the other. If she is tied up she will keep moving sharply over in her stall. She will turn her head to look towards her hind end (*photo 1*), and occasionally she may kick at her belly.

At the same time the wave of contraction, which extends throughout the entire uterus muscle, causes the so-called water-bladder, which surrounds the foetus, to press against and open up the cervix (entrance to the uterus) (*photo 3*). The precise mechanism of this 'opening up' or dilation is not fully understood, but it appears to be due to reflex stimulation, triggered off by the intermittent pressure and relaxation of the water-bladder caused by the uterine contractions.

These initial spasms of uneasiness occur every four or five minutes and last only for about three to five seconds.

What exactly is happening inside at this stage? Each time the muscular wall of the uterus (womb) contracts in labour (*see the model in photo 2*), the animal feels a slight, sharp pain and this produces her uneasiness.

As the first stage progresses, the uterine contractions become marked enough to cause the animal to arch her back and strain slightly (*photo 4*). The first strains usually occur at

2

intervals of three to four minutes and the actual strain lasts only for one second, though the back may remain arched and the tail cocked for between five and ten seconds.

Inside the patient the uterine contractions are now distinctly stronger and more frequent. The cervix is dilating progressively (*photo 5*).

6

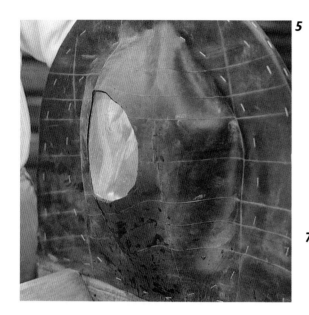

5

The cervix is now almost three-quarters dilated and the water-bladder is starting to balloon through the opening (*photo 7*). Up to this stage, dilation of the cervix is entirely dependent on the water-bladder pressure.

The strains will be seen to get stronger and stronger, and during the last few efforts before the end of the preliminary stage copious urine and dung are passed (*photo 8*).

7

Despite this increased straining, the patient will continue to behave normally, eating and drinking and being fully aware of everything around her.

However, there are two marked changes, both indicative of the pain of her labour. First of all, her breathing becomes much more rapid — about twice the normal rate — and, secondly, during the uterine contraction, the muscles of the brisket, neck and head region shiver and twitch markedly.

Towards the end of preliminary labour, the straining bouts become more frequent and each bout comprises several perceptible but mild strains.

All this while, the patient is standing up (*photo 6*) and/or walking round in alternate clockwise and anti-clockwise semi-circles. During the last hour of this first stage labour, the bouts of straining occur approximately every 1½ to 3 minutes and the number of perceptible strains in each varies from one to a dozen or more.

8

This is nature's way of making certain that as much room as possible is available for the passage of the calf down the vagina.

Finally, at the end of the first stage in labour, the pains of the uterine contractions make the cow lie down (*photo 9*).

An interesting and important feature here is that the feet are presented sideways. Further careful exploration will show that the calf is actually lying on its left side, with the left cheek of the head resting on the floor of the uterine body (*photo 11*).

9

11

This preliminary stage lasts for an average of two to three hours in a cow and four to six hours in a heifer, though I have seen it continue quite normally for very much longer than that. In humans the preliminary stage can last quite normally for up to 20 hours.

The cervix is now three-quarters dilated. A portion of the water-bladder is through into the anterior vagina and the feet of the calf are also poking through the cervix (*photo 10*).

The potential mother now goes into **second stage** labour which, as in humans, is a much more serious and intense affair.

The character of the patient changes markedly. Instead of being bright, lively and taking full notice of her surroundings, she appears to become oblivious and concentrates on her intense 'bearing down' (*photo 12*).

The intervals between the bouts of labour still vary from 1½ to 2½ to even 3½ minutes, but the bouts now comprise really vigorous strains each of which lasts from ½ to 1½ seconds.

10

12

4

After about a dozen of these severe strains of second stage labour, the calf has rotated to its normal correct horizontal position, i.e. it rotates through 90 degrees in an anti-clockwise direction. From now on, each strain causes the top of the calf's head to bear on the top inside of the cervix, and this renewed intermittent pressure causes the cervix to relax and open up further (*photo 13*).

On average, this stage is reached after seven or eight bouts of second stage labour strains — usually lasting about half an hour — or a total of between 30 and 40 strains, each strain lasting between ½ to 1 second. But with a big calf in a heifer, considerably more time and effort may quite normally be required.

Photo 15 shows this stage as it is inside the cow — head through the cervix and feet at the vulva.

13

15

The water-bladder travels down the vagina and the head of the calf starts to come through the cervix. At this stage the bladder may rupture, though it does not usually do so until the feet of the calf reach the vulva, and by that time the calf's head is through into the anterior vagina (*photo 14*).

The cow gets up and moves to a more comfortable position ready for the final efforts of delivery (*photos 16-19*).

16

14

17

19

18

20

After the appearance of the feet, the intervals between the bouts of straining become shorter, approximately every 15 to 90 seconds; the number of strains is greater and each one increases in intensity — now lasting from 1 to 2½ seconds.

Thirty to 40 of these strains usually see the calf's tongue appearing. At this stage the patient may rest for a minute or two, to allow the vulva to relax and to gain strength for her final effort. This, again, happens in women.

At least 50 or 60 strains, with varying rest periods of up to 90 seconds, are now required before the nose of the calf appears (*photo 20*). But again, with Friesian heifers particularly, normal progress may be much slower.

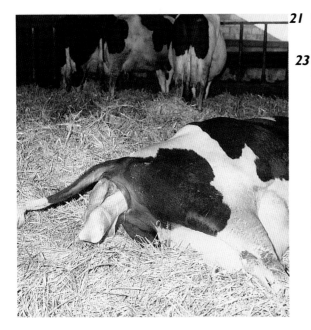

21

23

22

24

Another 50 strains will be required before the head appears (*photo 21*). During this 50, the rest periods are much shorter — from 15 to 60 seconds — and the strains are really intense and prolonged — lasting for up to 2½ seconds — just as though the animal knows that the labour is nearing completion.

After the head the rest is usually easy

(*photos 22-24*). With each strain, as the chest comes through, copious quantities of mucus may pour from the calf's mouth and nostrils. This is very important, since it is obviously nature's way of clearing the respiratory passages ready for normal breathing.

The live calf is finally delivered — unharmed in 99 cases out of 100 — ***provided the***

7

mother is left alone. But more of that later (*photo 25*).

In a heifer, this second stage takes on average three to six hours and in a cow, two to four hours. But with a big calf the second stage may continue quite normally for up to twelve hours or more.

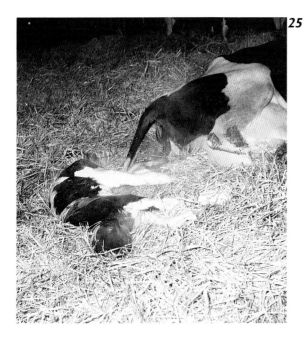

25

Within two or three minutes, the mother jumps to her feet and starts to lick the calf with a fierce maternal love (*photo 26*).

The calf is usually staggering onto its legs within 10 to 15 minutes.

Within 30 minutes, it has found the teats and is sucking down its precious quota of colostrum with its initial laxative and subsequent protective qualities.

I think all calves should be left to suckle their mothers for at least 24 hours but, of course, on many farms they are taken straight away at birth. I am convinced that leaving the calf with the mother is tremendously advantageous, firstly because the placenta and fluid on the calf contain hormones which play an important part in milk release and subsequent production. And secondly because, during the first four to six hours after birth, the calf gets, and can absorb and utilise, all the necessary antibodies against disease from the colostrum or first milk of the mother.

Now comes the **third or final stage** of bovine labour — the passing of the afterbirth (*photo 27*). Normally this occurs within an hour or two but occasionally it can hang for several hours. In such cases the retention is usually due to fatigue, perhaps caused by a big calf.

26

27

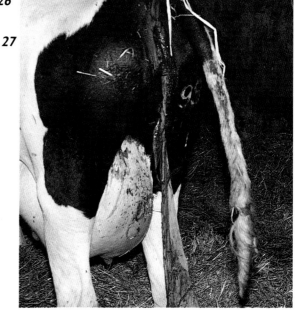

Where the afterbirth or cleansing is retained, then special precautions may have to be taken (see 'The Afterbirth', page 49).

Preparturient Oedema of the Vulva ('Hogging')

Occasionally a cow close to calving swells markedly around the vulva (*photo 28*) and in some cases the swelling extends down to the udder.

This condition, though alarming, does not normally interfere with the natural birth. Nonetheless, veterinary treatment is advisable.

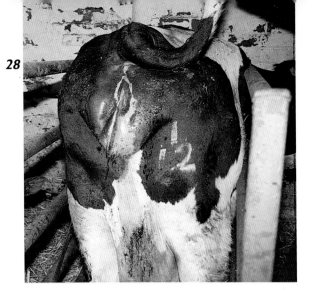

28

2
Dangers of Premature Interference

After studying natural birth, it must be obvious just how easy it is to retard nature's progress or to damage a heifer or cow by interfering during labour. For example, during the primary stage of labour, the cervix (the womb entrance) is being opened up by the intermittent pressure of the water-bladder. This cervical dilation is entirely dependent on the water-bladder pressure, especially the pressure exerted on the top part of the cervix. Any manual interference might lead to a premature rupture of the water-bladder and this causes a falling off of the intra-uterine pressure, the cervix ceases to dilate, and there is a serious set-back to nature's progress (*photo 1*).

I have seen this happen on many occasions, especially in the old days when premature interference was rife.

When the calf is finally presented in the

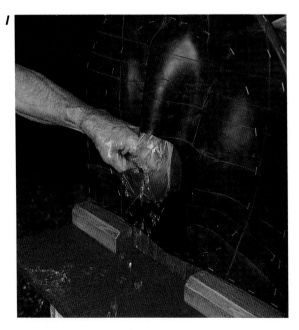

1

correct position, the cervix is rarely more than three-quarters fully dilated, even though the feet of the calf are now in the passage and the calf's nose may be peeping through. The continuation of the opening up of the cervix is now dependent to a very large extent on the intermittent pressure of the calf's head on the top part of the cervix.

Any interference at this stage will prove disastrous, since roping the feet and pulling forward will jam the calf in the cervix.

This cervix is composed of a powerful ring of muscles, and when the calf is jammed tightly, there occurs a spasm of the muscles — in other words, relaxation ceases alogether. This means that the mechanism of natural cervical dilation will no longer function. Further excess traction will rip the cervix and kill the animal (*photo 2*).

In fact, it is certain that manual interference at any time during the passing of the head through the cervix will prove disastrous. I have come across this cervical spasm caused by premature pulling on many, many occasions, and each time it has meant a prolonged embryotomy (i.e. cutting the calf up inside) or, in some cases, a caesarean section operation.

The next danger period is when the feet of the calf first come into view outside the vulva. In fact, I think this is the most important time of all, because many farmers rope and pull the feet as soon as they appear (*photo 3*).

This is all wrong. When the feet first appear, the cervix may still not be fully dilated, and certainly the vagina and vulva have not had a chance to relax to anything like their full extent. Forced traction or excess pulling at this stage, especially with a big calf, may still cause a ruptured cervix or, at the very least, will produce a torn and lacerated vagina and vulva (*photo 4*), with the ever present danger of fatal haemorrhage or secondary sepsis.

There can be little doubt that the best advice here is that, provided the calf is straight, and remember only if the calf is straight, leave the animal to complete the birth by herself. Have patience — the greatest gift of all at calving time. Don't panic so long as progress is being made.

The last thing you should be worrying about is the life of the calf; the calf is much more likely to die if it is pulled away than if it is left

2

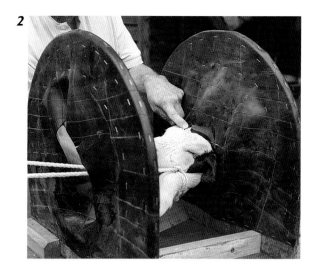

3

4

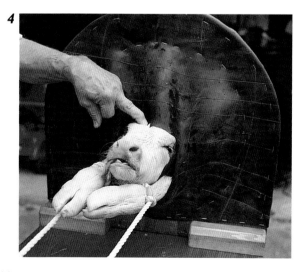

to come on its own.

The reason for this is quite simple. Whilst inside the cow, the calf receives its nourishment and oxygen from the mother via the navel cord. It does not start to use its own lungs for breathing until the navel cord ruptures, and this rarely occurs until the calf's head is well outside the cow. The intermittent straining of labour is designed by nature to maintain the navel supply of blood and oxygen, but prolonged premature pulling of the calf causes excessively long periods of pressure on the cord and this cuts off the calf's lifeline.

To sum up, therefore, the golden rule with all heifers or cows is to leave the animal to do the job by herself for at least a reasonable period of time, provided that the calf is coming in the correct position and that progress, no matter how slow, continues to be made (*photo 5*).

When the hind feet of the calf are coming first, i.e. in a posterior presentation, and when the hind feet first appear outside the cow, the cervix is usually only little more than half open. Excessive pulling on the legs at this stage will nearly always result in a ruptured cervix and a dead animal (*photo 6*). The correct technique for dealing with this is dealt with in 'Posterior Presentation', page 29.

5

6

3
When and How to Examine

Obviously it is important to know, with complete confidence, exactly when and how long to leave the patient and when to seek professional advice. The routine measures I advise are these.

If nothing is showing after a period of the intensive straining of second stage labour — a period of say two hours in a cow and four hours in a heifer — then examine for presentation.

11

First of all make the time to scrub the hands and arms thoroughly with soap, warm water and antiseptic (*photo 1*). It is surprising how few will take the trouble to do this and yet it is of great importance. I've seen many cows die of sepsis as a direct result of examination by a dirty hand.

Now wash the cow's vulva and perineal region — again a simple precaution often neglected (*photo 2*).

Next coat the hand and arm with soap flakes (the perfect lubricant) (*photo 3*) and wash the vulva thoroughly with flakes (*photo 4*).

Insert the hand very slowly and gently, taking great care not to rupture the water-bladder. It is best to strip off for the job because, apart from the fact that it is the only way to be sure of asepsis, you may well have to insert the arm right up to the shoulder. If

1

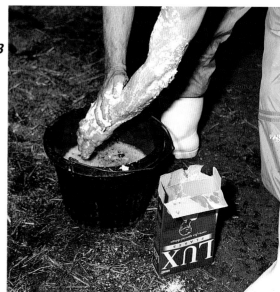

3

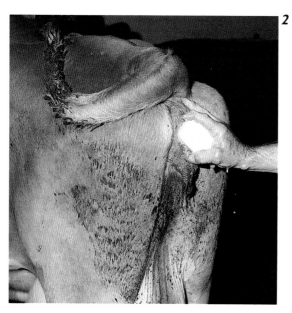

2

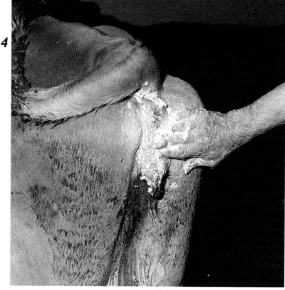

4

12

the calf's presentation is wrong, i.e. if the two fore feet and the head cannot be felt, then send for the veterinary surgeon at once (*photo 5*).

If the calf's head and forelegs are there, then relax and apply the following advice.

If it is spring or summer, turn the animal out with the others and leave her to get on with the job (*photo 6*).

If it is during the winter, arrange accommodation for calving animals according to the system of housing used.

Ideally put the patient in a loose box (*photo 7*) and cover the floor with sand or grit before bedding down (*photo 8*). **Bedding is not enough; the floor underneath must be completely non-skid.**

This gritting is a very good practical hint, since it is my experience that permanent paralysis (obturator and/or popliteal paralysis)

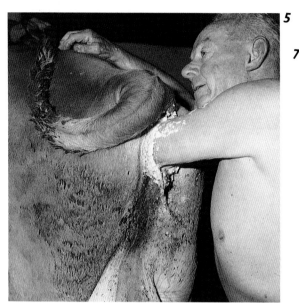

5

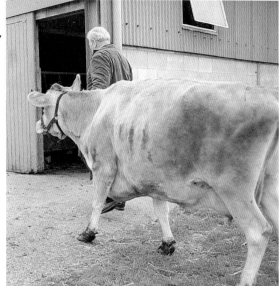

7

6

8

after calving is nearly always due to damaged or broken pelvises or hip joints directly caused by slipping about on a greasy floor during labour (*photo 9*).

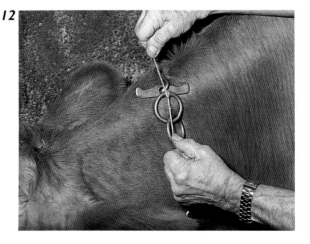

The obturator and popliteal nerves run down both hind legs and control muscle and joint movements. Excessive slipping of the leg, usually forward underneath the belly of the cow, damages these nerves in varying degrees and often results in up to a six-month convalescent period or a trip to the knacker's yard. (See 'The Downer Cow', page 74.)

In many cases this can be prevented by hobbling the patient either before or immediately after calving. Commercial hobbles are available (*photo 10*).

If the cows are kept in stalls and no loose box is available, accommodate the neighbouring cows elsewhere. This is simple commonsense, because the cow in labour needs some space to stretch out and a neighbour is more than likely to tread on her udder (*photo 11*).

Tie the patient carefully but loosely with a single link of string in the chain. This is a simple precaution against possible hanging which, I'm glad to say, the majority of herdsmen adopt instinctively (*photo 12*).

14

Just one point, though. Never use thick baling string; use a thin strand which is sure to break if the cow hangs back. I've seen several die because the baling string has not broken.

If, as in the majority of cases, the cattle are housed in cubicles or kennels, provide a non-skid yard deeply bedded with straw for all the dry cows and leave them there until at least 24 hours after calving.

If it is night time, the next thing to do is to go to bed and forget the patient. If it is morning, don't look at her again until late afternoon (see page 25).

4
General Hints on Assisting

TOOLS FOR THE JOB

For the Farmer and Stockman
In addition to the constantly required supply of powerful non-irritant antiseptic, the farmer or stockman should always have in his cupboard or medicine chest: three stout metal bars about two feet long; three nylon calving cords with a fixed noose on one or both ends; and a box of indispensable soap flakes (*photo 1*).

For the Veterinary Student
In addition to the basic requirements of the stockman, the veterinary surgeon should carry a sharp butcher's knife and steel (*photo 2*), not for butchering the cow as so many of my clients have jokingly inferred, but for the simple external embryotomy jobs like amputating the head of the calf or for cutting the foetus in half.

Another extremely valuable instrument for the veterinary surgeon is a self-closing hook

1

2

which can be used for bringing a head round or fixing any part of the foetus inside the cow (*photos 3 & 4*).

A spinal needle, syringe, local anaesthetic and sharp scissors are all vital for the administration of spinal anaesthesia as and when necessary (*photo 5*).

Wire cutters, embryotomy wire and flexible wire tubes are all essential for internal embryotomy, i.e. the cutting up of the foetus inside the cow (*photo 6*).

Here then we have the complete calving kit required by any veterinary surgeon to calve any cow (*photo 7*)

3

6

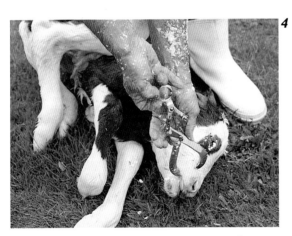

4

7

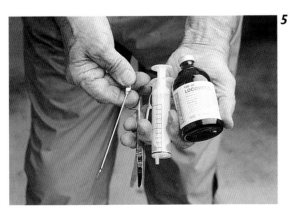

5

You will observe there are no barbaric eye hooks or weird and wonderful instruments. But above all there is no block and tackle, and no so-called calving machine. The skilled, intelligent veterinary surgeon will never use either of these under any circumstances.

16

HOBBLING AND GRITTING

One of the most common, if not the most common, causes of losses in first-calvers is damage to the hips, femurs or pelvis directly caused by slipping and sprawling about on a slippery floor during calving.

Even though the patient is lying comfortably to begin with, she only has to attempt to get up, and in so doing splay her legs once, for irrevocable damage to be done — damage which will mean she will finish up in the knachers yard.

This is especially the case with heavier breeds, and I would say that splaying occurs most frequently in Friesian heifers, especially when the calf is by a Friesian bull.

To me these losses are almost criminal, because they can so easily be avoided by following a simple routine.

First of all, clear the bedding from underneath the cow's hind feet and sprinkle the entire area in front of, between, at the sides and at the back of the legs with sand or grit (*photo 8*). It's **never** enough just to put plenty of bedding down because the restless animal soon pushes the bedding aside. The floor underneath is often as slippery as ice, perhaps because of milk running from the udder.

Now hobble the animal. If a commercial hobble is not available take two short lengths of good strong rope (in an emergency several strands of strong binder twine will do), and tie the first length above the fetlock in a reef knot (*photo 9*).

9
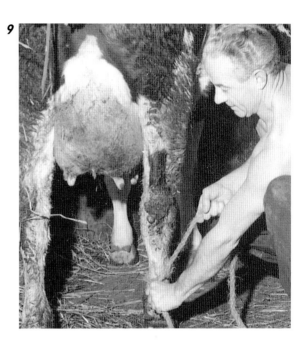

Tie the other piece of rope similarly round the opposite leg and join the two pieces of rope in the centre, again by a tight reef knot. If there are any excessively long loose ends, cut them off with a knife. The legs should be approximately 45 cm apart (*photo 10*).

8

10

Having fixed the hobble, spread the rest of the sand or grit over as much of the bed area as it will cover.

Now you can get on with the job with all the confidence in the world and the full knowledge that you may have saved your animal by taking these simple basic precautions. Don't worry about her. She can get up and down with the hobbles fitted and can, if necessary, walk around for days or weeks with them on.

LUBRICATION

Many lubricants have been used in bovine obstetrics. Liquid paraffin, linseed oil and lard were all favoured by many of the older generation of our profession.

I have found, however, that the best lubricant of all is probably the simplest and most easily obtained — soap flakes. Copious quantities of soap flakes and plenty of hot water (*photo 11*) produce a lather and lubrication which at times is little short of miraculous.

Looking back over 53 years of tough practical experience, I can think of many cases where soap flakes have made all the difference between success and failure. I am now so firmly convinced of the value of soap flakes that I buy them by the hundredweight.

I have often heard it said that an ounce of lubrication is worth a ton of pressure. In obstetrics there is much truth in this and I cannot stress too strongly the inestimable value of the lubricant qualities of soap flakes in all cases of difficult or prolonged parturition (*photo 12*).

And I would exhort every veterinary student, in particular, to pay great heed to this simple hint. It will prevent many a death and many a frustrating heartache. I only wish someone had told me about soap flakes at the start of my career.

12

11

ROPING A FOOT AND FIXING THE CALVING ROPE TO A METAL BAR

Assistance at calving should only be given when the animal has ceased to make natural progress. More specific details of when and how to assist are given in later chapters but here, first of all, I illustrate the simple processes of roping a foot, and of fixing the rope correctly to a bar.

These simple hints are important because:

● If a calving rope is tied too low down towards the calf's foot, the claw may come off when pulling.

• An incorrectly tied knot on the bar is not only liable to slip persistently during the job, but is also often difficult or impossible to untie afterwards. The knot which I illustrate is easily untied, even after the strongest pressure.

Wash the cow's vulva region thoroughly with soap, water and a reliable non-irritant antiseptic; then thoroughly wash your hands and arms. I cannot emphasise too strongly the ever present need for absolute cleanliness during assistance at calving.

Make a running noose (*photo 13*) on the end of a calving rope which has either been boiled or soaked in a strong solution of non-irritant antiseptic.

14

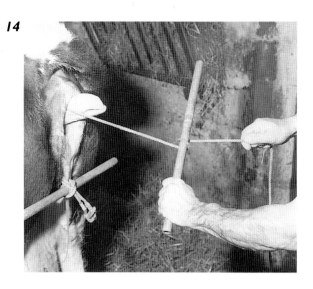

13

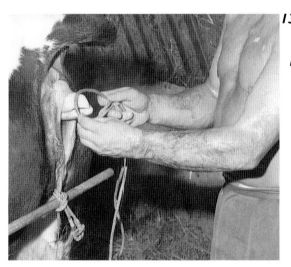

15

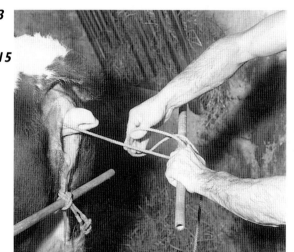

Fix the noose **above** the calf's fetlock. If the rope is allowed to slip down to the calf's coronet it is liable to come off, pulling the horn of one of the claws with it.

Having fixed the noose, lay the bar on top of the rope and pass the end of the rope over and round the bar to form a half hitch (*photo 14*).

Then take a loop of the free end of the rope forward and pass it underneath the main part (*photo 15*).

Bring the centre of the loop through the larger loop thus formed, by pulling it with the thumb and forefinger exactly as shown (*photo 16*).

16

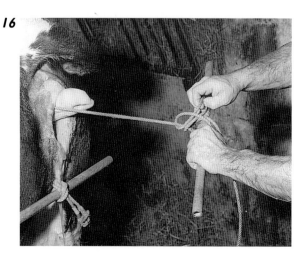

Holding the end of the loop rigidly by the forefinger, pull the bar towards you until the knot tightens (*photo 17*). The bar is now fixed by a knot which will stand any amount of pressure but which is comparatively easy to untie.

new set of ropes will be needed for each calving case. Provided the knot has been tied as illustrated earlier, there is a simple technique for undoing it.

Drop the bar onto the floor, retaining a hold on the free end of the rope, i.e. the end opposite to the one that has been attached to the calf (*photo 19*).

Place the left foot firmly on the bar on one side of the knot (*photo 20*).

17

RELEASING THE ROPE FROM THE BAR

A simple but common problem after assisting a cow to calve is that of undoing the knot on the bar (*photo 18*). Even with controlled pulling, the average pair of hefty shoulders can pull the knot remarkably tight.

With nylon calving ropes costing a fair amount of money, it is important that the rope should not have to be cut; otherwise a

19

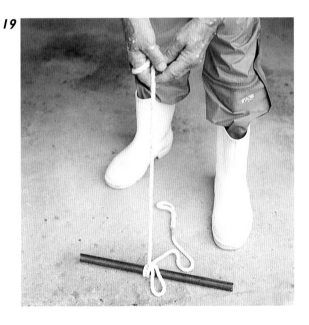

20

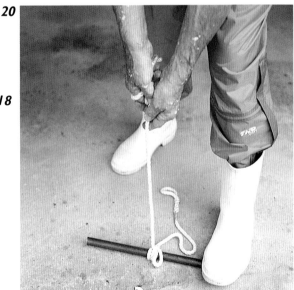

18

Bring the right foot onto the other side of the bar and bear all your weight evenly on both feet. Then wrap the free end of the rope several times round the hand (*photo 21*). If the knot is very tight, it's a good idea to protect the hand with a towel or piece of sacking before wrapping the rope around it.

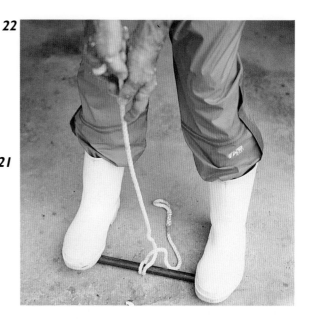

22

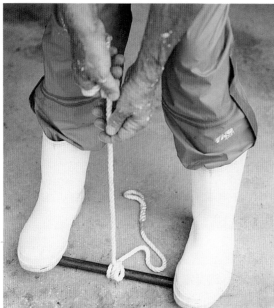

21

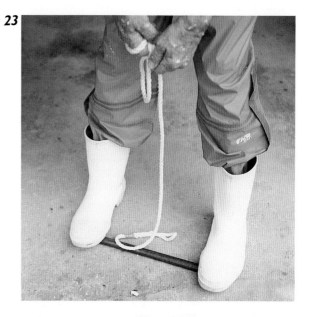

23

Using two hands, pull with all your strength. A tight knot will require considerable upward pulling, but I have yet to see one which will not untie by this method (*photos 22 & 23*).

HOW TO ASSIST WHEN THE COW OR HEIFER IS STANDING

One of the most difficult tasks is to assist a cow when she is calving standing up. A heifer, particularly, is often unwilling to lie down during interference; in fact, to get her to lie down, casting is sometimes necessary.

The trouble is that, when pulling by the normal method is tried, it is virtually impossible to exert any effective pressure, and the problem is often complicated by the persistent restless movement of the patient. Here I show an original idea which I have found effective on many occasions (*photo 24*).

24

Having fixed the calving rope to a bar, drape the bar, on both sides of the rope, with a sack or towel (*photo 25*).

Holding the bar rigidly in position with the right hand, pass the left leg over the other side of the bar. The bar tension can now be taken equally by the backs of both thighs (*photo 27*).

25

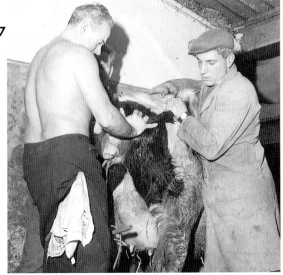

27

Now pass one end of the bar underneath the right thigh (*photo 26*).

Now place both your hands one on either side of the vulva and, as the cow strains, push backwards, taking the full weight of the pull on the back of the thighs (*photo 28*). The towel or sack helps to cushion the bar and also keeps the trousers reasonably clean.

26

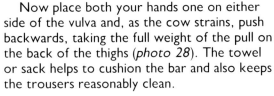

28

As progress is made, pressure can be varied in direction by using one arm only and pulling to the one side (*photo 29*).

By using this method, at least twice as much power can be obtained as that possible when using the traditional method and, even more important, the fatigue is negligible compared with that associated with arm pulling.

It is very important at all times, whether the cow is standing or lying, to exert pressure only when the animal strains and to relax completely when she relaxes. The old idea of maintaining a steady pressure during assistance is entirely wrong, because steady pressure causes spasm of the uterine neck muscles and holds up the entire process.

It is very important to know about this 'deficiency delay' because any attempt to pull the calf away from such patients can, and often does, result in a prolapsed womb and a dead cow.

The first step is to see how the calf lies. This examination in 'delay' cases is really a job for your veterinary surgeon and one you shouldn't tackle unless you are very experienced and quite unable to get professional help. The reason this examination is so much trickier than in straightforward labour is that usually the calf is right inside the womb and the cervix may only be partially dilated.

Having scrubbed up and lubricated thoroughly, the veterinary surgeon will insert his hand very carefully, taking great care not to rupture the water-bladder. He will examine for the calf's two fore feet and head. If they are there, then the calf is straight and properly presented and the cow will most assuredly calve it herself.

Once it is established that the calf is presented correctly, you may be almost sure that all that is needed is 453g (16 oz) of calcium borogluconate solution. This should be injected underneath the skin, approximately a hand's-breadth behind the central ridge of the scapula (main bone of the shoulder joint) (*photo 30*).

29

DEFICIENCY DELAY

Often I am called out to attend cows coming down with their third or fourth calf. The history usually is that they have been shaping for calving for anything up to 16 hours but are not getting on with the job.

In such cases the trouble, almost invariably, is caused by calcium deficiency.

30
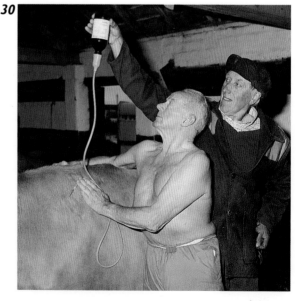

The correct technique is not so easy as it looks and this is another reason why the treatment should only be given by a veterinary surgeon.

Another essential feature of the calcium injection is a thorough rubbing away of the injection swelling. This is essential not only to ensure rapid absorption of the solution, but also to avoid abscesses and permanently unsightly lumps (*photo 31*).

Calcium injected subcutaneously takes approximately 20 minutes to take effect. At the end of that time the cow will go into normal powerful second stage labour and will deliver the calf by herself.

If it is summertime the cow can be turned out to calve at pasture (*photo 32*). If it is winter, then follow the instructions given on pages 13-15.

In all such cases of protracted labour due to calcium deficiency, the cow should be left to calve herself after the calcium administration. Calcium deficiency causes a loss of muscular control in the neck of the uterus and, consequently, if the calf is pulled out, there is great danger that the womb will prolapse behind it.

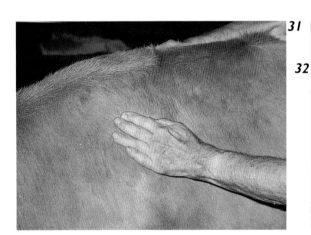

31
32

5
When and How to Assist

The cow has been examined, the calf has been found to be presented correctly (i.e. two fore feet and head coming), and the patient has been moved as necessary to a suitable place for calving, and has been provided with a non-skid floor or has been hobbled on a gritted floor. The next obvious question is just how long such a case can be safely left before doing something about it.

It is difficult to generalise but,

provided the foetus is presented correctly (and I cannot stress this point too often), the patient can be left with complete safety for a further eight to twelve hours (photo 1).

In other words, if calving is not completed by the following morning (if you have left the patient over-night) or by late afternoon (if the examination was done in the morning), then the veterinary surgeon should be called in. He will decide whether to assist or whether to leave her to her own devices for a further period, depending on the size of the calf, progress being made, and the general state of cervical, vaginal and vulval relaxation.

However, there are occasions when a keen practical farmer or stockman can detect the cessation of progress in the birth and can intelligently assist by himself, or at least attempt to do so, before consulting his veterinary surgeon, or perhaps more important, know how to assist if the veterinary surgeon is not immediately available. I am thinking particularly of three fairly common situations: when the calf's nose has been showing for some time and the head and tongue are swelling markedly; when the calf is apparently stuck at the hips; and when the hind feet are coming first.

Before dealing with these situations, however, I want to emphasise probably the most important point of all in the calving of heifers or cows.

Excess force should never be used. In the vast majority of cases no more than one man, and one man only, should be allowed to pull on the ropes, and then only when the animal strains. In the few instances where two people are allowed to pull, they must do so only under strict veterinary supervision. I have proved many times that, in the odd cases where nature fails (and provided you wait until natural progress ceases), one man with the lubrication of soap flakes and lots of patience can calve the tightest and most difficult case (photo 2).

The only time any form of excess traction is permissible is when the calf is dead and putrefied, because then the rotten calf will stretch before the cervix tears or the vagina and vulva rupture.

NOSE PRESENTED WITH HEAD AND TONGUE SWOLLEN

When the nose of the calf has been showing for three hours or longer and no further progress has been made, then assistance is required because, after that length of time in such a position, the calf's head and tongue may become markedly swollen (*photo 3*).

The tendency in such cases is to rely entirely on pulling on the feet of the calf. Although it is often necessary to pull on the legs alternately, this in itself does not usually work, because it merely tends to bring both shoulders forward and increase the impaction. I have found the best procedure to be as follows.

Having first sanded or gritted the floor underneath the bedding and hobbled the animal as described previously, thoroughly wash the hands and arms as well as the animal's vulva and perineal region. Then take several handfuls of soap flakes and with the hot water lubricate between the top of the calf's head and the top of the vagina (*photo 4*).

Next, take one of the nylon calving ropes which has either been boiled or soaked in a strong solution of non-irritant antiseptic and make it into a loop (*photo 5*).

Insert the centre part of the loop over the top of the calf's head to just behind the ears (*photo 6*), keeping the free ends of the rope outside the vulva (*photo 7*).

6
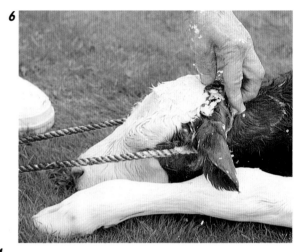

4
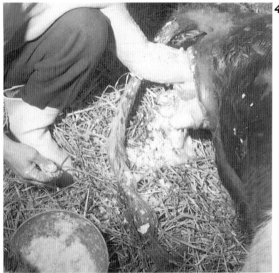

7

5

Tie the ends to a bar and then, as the cow strains (and only as she strains), moderate pressure on the bar will soon bring the head forward.

It will probably be necessary to rope and bar both the forelegs, and to pull on these in turn with the head rope (*photo 8*), before the head finally emerges from the vulva. Once the head of the foetus is delivered, the rest of the job is usually easy.

During assistance, a bale of straw placed

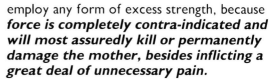

8 employ any form of excess strength, because *force is completely contra-indicated and will most assuredly kill or permanently damage the mother, besides inflicting a great deal of unnecessary pain.*

It is likely to be a case for the veterinary surgeon, but several emergency first-aid methods should always be tried before sending for the vet.

First of all, turn the cow onto her back, then over onto the side opposite to the one you found her on (*photo 10*). Leave her completely alone for five minutes; then one person should try gentle assistance as and when she strains. This simple procedure, and only this, will release 50 per cent of calves in such cases.

between the animal's hind end and one of your feet is useful. This not only prevents the cow moving backwards, but allows a single puller to exert much more effective pressure during the animal's strains.

STUCK AT THE HIPS

The next presentation where interference is necessary is where the foetus apparently gets stuck at the hips (*photo 9*). This occurs fairly commonly, especially in Friesian heifers.

In such cases there is always the feeling that a little more help is all that is required and the farmer often seeks additional help, finishing up pulling the cow round and round the box and sometimes even round the outside yard. Never be tempted to do this, or for that matter to

10

If this fails, rope the forelegs of the foetus together using a figure of eight knot. Tie the **9** knot as tight as possible.

Insert a bar between the tied legs and slip the bar up to the calf's elbows (*photo 11*).

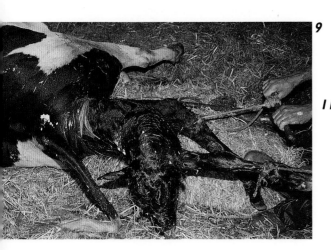

11
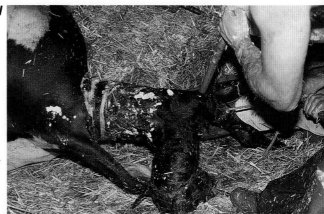

With someone pulling on the ropes, and again only as the patient strains, rotate the foetus on its own axis — first in one direction and then in the other. This will release most of the other cases (*photo 12*).

13

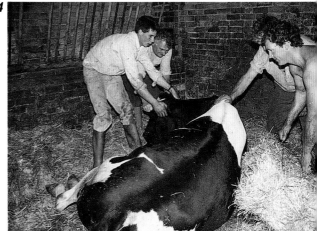

12

Two other tips might be tried.

First of all, pass a strong sack underneath the body of the calf and as close to the vulva as possible. Gripping both ends of the sack firmly, pull as strongly as possible at right angles to the long axis of the mother. At the same time, get the assistant to pull downwards on the legs. The idea here is to assist the stifles of the foetus over the brim of the mother's pelvis and I have used this successfully on a number of occasions (*photo 13*).

Alternatively, and based on the same principle, push one of your shoulders underneath the diaphragm of the calf and lever outwards as the assistant pulls downward in the opposite direction. In this case it is worth trying to twist the foetus simultaneously, first in one direction and then in the other.

If all these methods fail, send for your veterinary surgeon. He will probably cut the calf away in the manner described in 'Embryotomy', page 32.

Always, immediately after a delivery where the patient is prostrate, set her up on her brisket to stop her straining (*photo 14*).

Note the golden rule: never have more than two men assisting at any one time; excess strength is always contra-indicated (*photo 15*).

14

15

POSTERIOR PRESENTATION

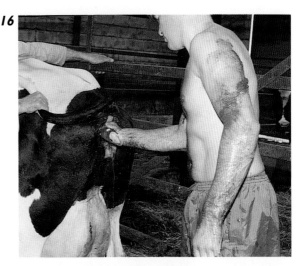

16

When a calf is being born hind feet first, the common idea is that it suffocates by breathing in the fluids from the uterus. This is not so, at least not until the calf is nearly born, because so long as the navel cord remains intact, the calf is kept alive by blood which flows in through the cord from the mother.

The calf does not use its own lungs for breathing until the navel cord ruptures, and this does not occur until the calf is nearly born — certainly not before the entire hind end of the calf is outside the vulva.

The following series of photographs show how to calve a **live** calf in a posterior presentation; if the instructions are followed carefully, a dead calf in these circumstances will be a comparative rarity.

17

Usually when the calf is coming backwards, the cow takes longer than normal to get the feet out and she doesn't appear to strain anything like as hard as when the calf's fore feet and head are presented correctly. This is because, in the posterior presentation, there is no head to exert pressure on the top part of the cervix after the rupture of the water-bladder (*photo 16*).

How do you tell when the calf is coming hind feet first? Simply by the fact that the claws, or feet, are upside down (*photo 17*). It is surprising how often some stockmen forget this elementary rule of diagnosis.

Nonetheless, it is wise to make sure that they are hind feet, because the calf may be coming upside down. Scrub up thoroughly and examine for the hocks and tail (*photo 18*).

18

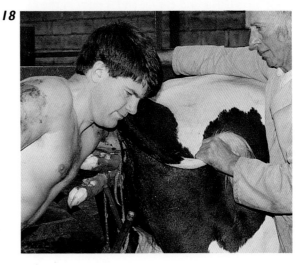

At this stage there is no need to panic. In fact, it is quite safe and often wise to allow the cow two, three or even four hours of normal straining after the hind feet appear.

This gives the cervix time to relax and open up completely. Because there is no head to exert pressure on the top part of the cervix, it takes longer to open. Consequently, premature pulling can be disastrous, as illustrated on the model in 'Dangers of Premature Interference', page 2.

Obviously, therefore, it is much better to have a veterinary surgeon supervise the birth. A veterinary surgeon called in the early stages of a posterior presentation will often inject a muscle relaxant intravenously and leave the patient for an hour or so before proceeding to help.

Having confirmed that the feet are hind feet, and having waited until the cervix has opened up, rope both feet above the fetlocks and fix the ropes to bars as instructed previously.

One person, and only one person, should pull on the bars, pulling one leg at a time and only when the animal strains (*photo 19*).

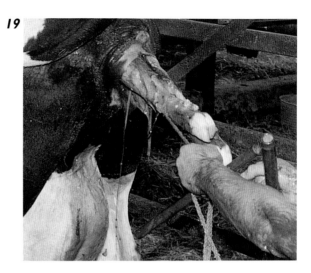

Even with only one person pulling, when pressure is applied the patient will often go down. When this happens, release the pressure immediately so that the cow does not sprawl awkwardly. If the patient is a heifer, of course, the legs should be hobbled and sprawling should be a lot less likely.

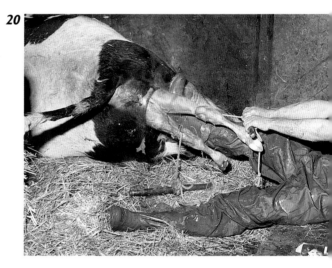

If and when the cow does go down, always make her comfortable by pulling both hind legs out from underneath her. Needless to say, the job of assisting is very much easier with the cow lying down (*photo 20*).

Often when the calf's hind end comes into the vagina the cow will get to her feet or attempt to do so. It is wise to allow her to stand and/or relax for a few minutes if she wishes. This is nature's way of helping delivery by changing the area of abdominal pressure on the calf.

When the calf's hocks appear outside the cow's vulva, it usually means that the entire hind end of the calf has passed through the mother's pelvis and cervix and is now in the vagina. When this stage is reached, the rest of the job is usually comparatively easy (*photo 21*).

At this stage, the pulling should still be confined to one person, persisting with one leg at a time and only as the cow strains. The reason why it is essential to maintain this intermittent pulling is that blood must be allowed to flow freely through the navel cord; otherwise the calf will certainly die. When the cow strains, the flow is interrupted for a few seconds by the pressure on the cord between the cow's pelvis and the calf's abdomen or chest. But when the cow relaxes between each strain the precious blood supply is resumed.

The old traditional idea of maintaining a steady pull is absolutely wrong and has been responsible for the loss of countless calves. The 'steady pull' idea should be banished from stockmen's minds now and forever.

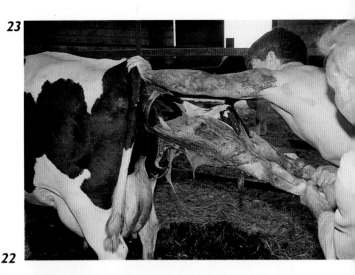

23

Up to now assistance should have been casual, calm and completely unhurried. But, and this is the most important point of the whole series, **as soon as the entire hind end of the calf is outside the vulva it is a matter of urgency that the calf is delivered quickly** (*photo 22*).

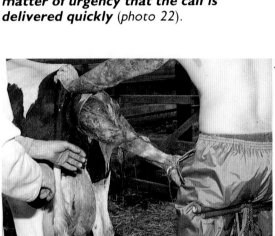

22

24

At this stage, the navel cord is trapped tightly between the cow's pelvis and the lower part of the calf's chest, and the life-giving blood supply from the mother is cut off. The cord may rupture at any moment. When this happens the calf will involuntarily breathe in and fill its lungs with the uterine fluids.

Consequently, as soon as the hind end of the calf is out, get several men to help and pull like hell. You have approximately 30 to 40 seconds in which to deliver a live calf. You won't injure the cow at this stage, because the widest part of the calf's chest is now through the mother's pelvis and cervix (*photo 23*).

If the calf is alive when assistance is started, it will have a 99 per cent chance of survival if this routine is adhered to (*photos 24 & 25*).

One inflexible rule, especially for veterinary surgeons and students, is that after a difficult calving case and no matter how protracted the

25

abnormal presentation may have been, always check for the possibility of a second calf (*photo 26*). Failure to do so can not only damage a reputation but can cost the life of the cow.

EMBRYOTOMY

Embryotomy simply means the cutting up of a dead foetus while the foetus is still inside the cow, whether the foetus is lying wholly or partially in the uterus.

This, of course, is a job for a veterinary surgeon. However, knowledge of the method employed is not only interesting, but it should assist agricultural students and farmers to understand better the tasks of the veterinary surgeon. It should also highlight the tremendous advantages of embryotomy over forced traction.

When animals are interfered with prematurely, a spasm of the cervical muscles occurs and the foetus jams in an incompletely opened cervix. When this happens, embryotomy is not only essential but it becomes arduous and complicated — legs have to be peeled off, the body of the foetus has to be whittled down in size, etc.

However, by allowing the animal to progress naturally, the need for extensive complicated embryotomy disappears completely. In fact, if nature is given a reasonable chance, embryotomy is required only for the following.

- Amputation of a turned-back head when the foetus is emphysematous (i.e. when it is dead and blowing up with putrefying gases).
- Very occasionally in a breach presentation (i.e. when the calf's tail is coming first), when the leg joints are stiffened (ankylosed) and it is not possible to bend the hock forward or the fetlock backward.
- When the foetus gets stuck with its hips in the pelvis of the mother.
- In a schistosomus reflexus (i.e. a monstrosity where the foetus is turned inside out), when the rudimentary legs are pointing downwards into the uterus. When the rudimentary legs are presented, especially in a heifer, then a caesarean section is usually indicated.

Exactly the same embryotomy technique is used for the amputation of the turned-back head of the foetus, the cutting off of an ankylosed hind leg, the removal of a foetus which is stuck at the hips, and the amputation of portions of the schistosomus. In each case the job can be done quickly and efficiently, without any damage whatsoever to the mother. The photographs and text describe embryotomy carried out on a foetus stuck at the hips.

First of all, using a sharp butcher's knife, cut the foetus in half as close to the patient's vulva as possible (*photos 27 & 28*). (A spinal anaesthetic may have to be administered, depending on the duration of the labour and degree of exhaustion of the patient.)

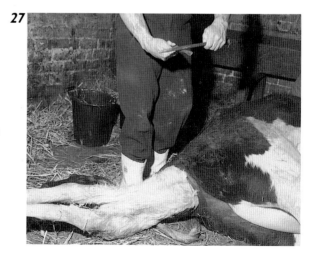

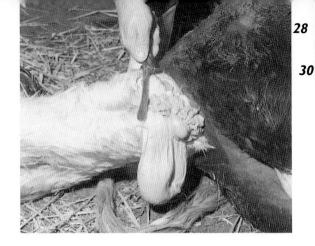

28

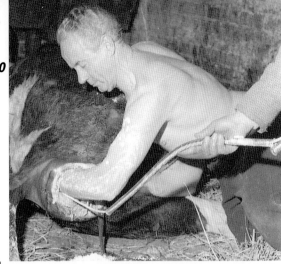

30

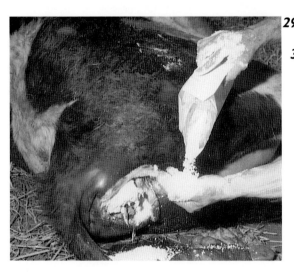

29

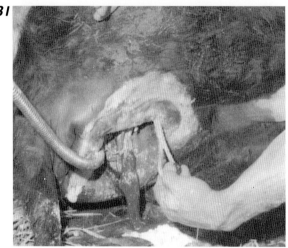

31

Next, cover the stump with a copious quantity of sulpha powder (*photo 29*). This is a simple precaution but well worth doing as a routine. The sulpha drug helps to control any infection over and around the vulva area.

32

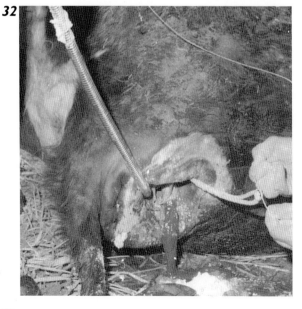

Thread a calving rope through one of the flexible wire tubes, and pass the looped end of the calving rope into the uterus and over the top of the tail of the foetus (*photo 30*).

Pass the hand around the hind end of the foetus to between the two stifle joints. Feel for the cord loop and, when you find it, pull it to just outside the vulva, having simultaneously guided the metal spring along the vagina to the foetal tail region to protect the delicate tissues of the mother from the rough surface of the wire (*photo 31*).

Tie a good length of embryotomy wire to the loop. A well tied reef knot is essential here, even though the wire is not easy to handle (*photo 32*).

Thread the wire through the second wire tube. It is usually necessary to use a second calving cord to do this, i.e. pass the cord through the spring, tie the other end of the wire to the loop of the cord, and then pull it through (*photo 33*).

34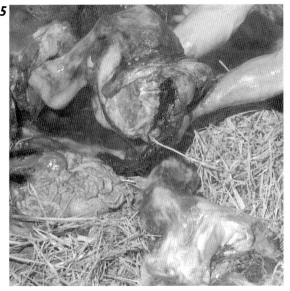

33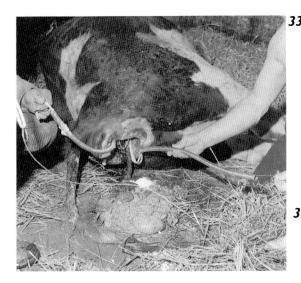

35

Instead of halving the foetus completely, the foetal abdomen can be transverse sectioned and its contents removed. Such a technique ensures against the amputated stump sinking back into the uterus and out of reach. Such a happening is rare but nonetheless can occur.

Another useful practical hint is to attach a weight (such as one of the metal handles used with calving chains) to the loop end of the rope that is passed over the foetal hindquarters. The weight makes the search for the loop end considerably easier.

Cup the hand over the junction of the wire and cord and, as the assistant pulls the cord through the first metal spring, guide the second spring into position between the thighs of the foetus. Tie each end of the wire to a bar, cross the springs over each other, and start to saw (*photo 34*).

It takes only a few seconds to saw through the pelvis of the foetus and the hindquarters can then be easily removed in two halves (*photo 35*). The mother is completely undamaged and, within minutes of the operation, will sit up and start to eat or drink.

A slight variation of the technique is illustrated in colour photos 36-43. It has been evolved because of the difficulty in purchasing replacement metal springs.

The tremendous advantage of this simple embryotomy over the force extraction by calving implements, gangs of men, a block and tackle or tractor is self evident and crystal clear. The illustration of this fact is probably my main reason for including this section on embryotomy.

A large calf stuck at the hips and immovable despite rolling and twisting, etc.

A rope is passed over the top of the calf's tail and down between the inside of the calf's stifle joints.

36

38

Using a sharp knife, cut through approximately 95 per cent of the dead calf leaving the spine intact as an insurance against the calf's pelvis disappearing back out of reach of the veterinary surgeon.

The embryotomy wire is attached to the rope and pulled through into position.

39

37

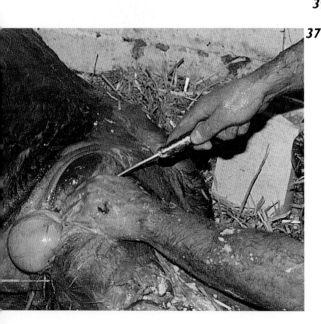

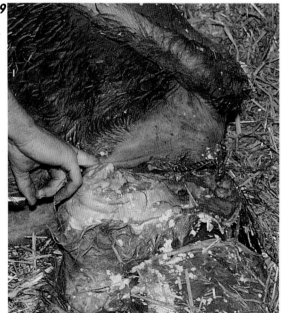

The wire is attached to a handle and one bar.

The sawing of the foetus is almost complete.

40

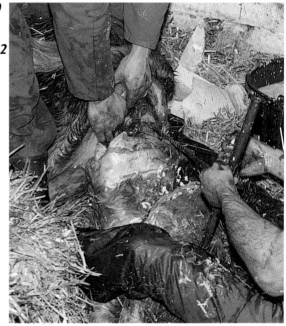

42

Since no metal springs are available, the patient's vulva and vagina are protected by first of all crossing the wire outside the vulva, and secondly by getting an assistant to hold back as much of the vulva as possible after soaking his hands in antiseptic. Then sawing commences.

The two halves of the foetal pelvis are easily removed.

43

41

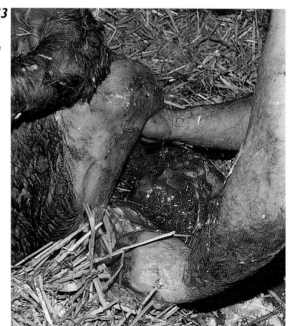

6
Simple Malpresentations

Only a qualified veterinary surgeon should attempt to correct a malpresentation. Even a comparatively inexperienced qualified assistant in a country veterinary practice has had more experience of malpresentations than the average farmer or herdsman.

This is simply because trouble at calving time on the individual farm is comparatively uncommon whereas in a veterinary practice trouble is our business. In fact, the average veterinary surgeon rarely sees a normal calving.

However, in an emergency, the veterinary surgeon is not always immediately available and, if the malpresentation is a simple one, then an intelligent herdsman should be able to cope. In any case, the techniques described in this section should be of great interest to all stockmen and farmers and should prove of considerable value to veterinary students.

A first essential, before any form of examination or interference is attempted, is an abundance of hot water and plenty of powerful though non-irritant antiseptic (*photo 1*).

If the cow is in a loose box, secure her head by tying her up with a halter. This may seem elementary but it's surprising how often examination and assistance are attempted with the patient continually moving around.

As suggested earlier, it is much better to strip off for all calving cases. It is the only way to attain complete cleanliness and it allows one to insert the arm up over the shoulder joint. Having thoroughly washed the cow's vulva and the hand and arm, examine for presentation.

HEAD TURNED RIGHT BACK

This is a common malpresentation and, although its correction is comparatively easy in a cow, in a heifer the job can be extremely difficult.

It is necessary to get right inside to assess the calf's position (*photo 2*). In the case illustrated, the calf's head was turned sharply back along the left-hand side of the uterus.

To set about rectifying such a presentation,

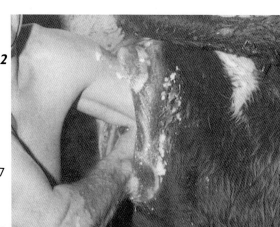

first of all place the flat of the hand on the calf's brisket and push it back into the uterus in the intervals between the cow's strains (*photos 3 & 4*).

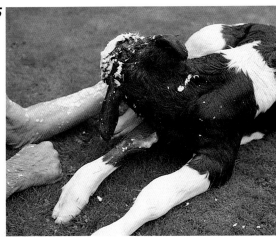

5

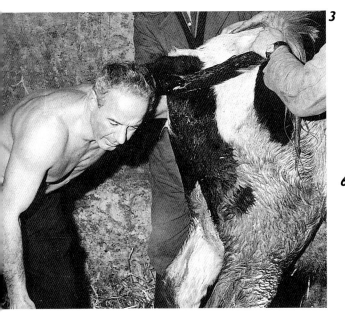

3

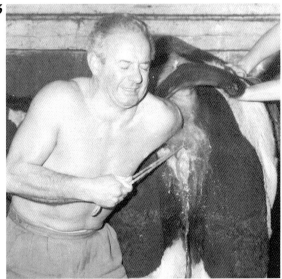

6

4

With the calf completely back in the uterus, pass the left hand (or right hand if you are right-handed) along from the calf's ear to underneath the jaw (*photo 5*).

Turn your body away from the cow so that your back is facing her head and, in this position, lever the calf's head round (*photo 6*). Of course, you have to be fit and fairly strong for this job. Photo 7 shows exactly what is being done inside the cow.

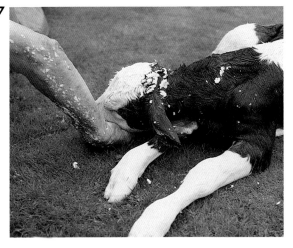

7

In the vast majority of cases in cows, as soon as the head is straightened the calf will be pushed out rapidly. However, in heifers it is often only possible to partially straighten the head, owing to the lack of sufficient room. This causes an S-shaped bend which is dealt with in the next section.

S-SHAPED BEND IN NECK

Often in a heifer, and occasionally in a cow where the calf is very large, it is difficult or impossible to straighten out by hand a head which has turned right back.

What happens is that, although one can bring the calf's nose round to the entrance of the cervix (i.e. so that it rests on the cow's pelvis), there is an S-shaped bend in the calf's neck. Each time the cow strains, or if pressure is exerted on the calf's feet, the head stays where it is or twists back again.

There is a simple method of overcoming this.

First of all, take a calving rope which has either been boiled or soaked for several minutes in a strong solution of non-irritant antiseptic (*photo 8*).

Having thoroughly washed and lubricated the vulva and the hands and arms insert the loop of rope into the vagina (*photo 9*).

9

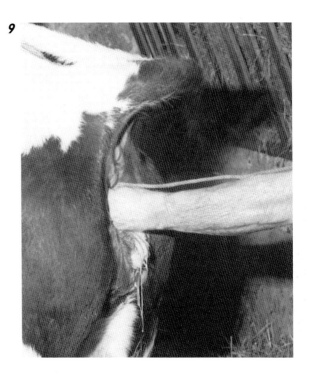

8

Pass it along over the top of the calf's head to just behind the ears, maintaining moderate pressure with your other hand on both free ends of the rope all the while (*photo 10*). You will have to put the loop over one ear at a

10

time and, unless a steady pressure is maintained, the loop will slip off the first ear as you are adjusting it over the second (*photo 11*).

Each time the cow relaxes after straining, exert moderate pressure on the rope around the calf's head. You will find that, in 99 cases out of 100, the head will shoot forward into its correct position. This is one of the few times when you do not pull as the animal strains (*photo 13*).

11

13

It is not necessary, as so many people think, to pass the ends of the rope into and through the calf's mouth. In fact, if you do this, you will often irreparably damage the calf's jaw and will certainly jeopardise the calf's chances of survival.

Attach the two ends of the rope to a bar (*photo 12*), as instructed on pages 18-20, and attach other ropes and bars to both the calf's fore feet.

HEAD OUT AND LEG BACK

A reasonably common malpresentation, especially in heifers, is where the head and one fore leg, or the head alone, is outside the vulva. Frequently, when found, the head and tongue are markedly swollen (*photo 14*).

12

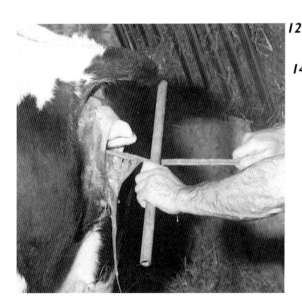

14

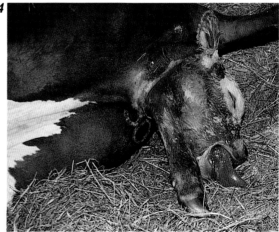

To attempt to pull the calf away in such a situation will, in a heifer, prove difficult or impossible, or at the very least will rip and injure the heifer, and at the most will kill her. If the calf is alive, then most assuredly it is a job for your veterinary surgeon (see page 52).

However, if the calf is dead, the head excessively swollen, the heifer in acute distress, and a veterinary surgeon not immediately available, then it is possible, on occasion, for an intelligent stockman to manage.

In any case, the correct technique with a dead calf is as follows.

Having first of all hobbled the cow to prevent splaying, pass the running noose of a nylon calving rope over the top of the calf's head to behind the ears and then tighten the noose (*photo 15*).

16
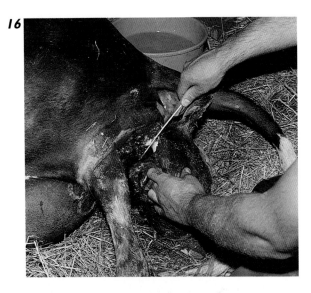

17

15
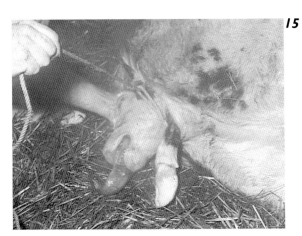

With one man pulling firmly on the rope, cut the calf's head off just behind the ears. Use a sharp butcher's knife or bread knife, taking great care not to cut the vulva. It's a good idea to use one hand to hold the membrane and skin of the vulva away from the cutting area during the operation. If an assistant is not available, extra care is necessary (*photo 16*).

Now push the whole foetus back into the uterus, pushing only between the cow's strains, and exerting the main pressure with the hand over the stump of the neck (*photo 17*).

Now search for and find the bent back fore

foot, and introduce the looped end of a calving rope. Pass the noose over the turned back foot, this time to just above the hoof, and tighten it around the calf's pastern (*photo 18*).

18
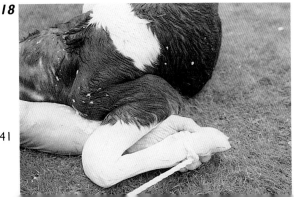

41

Pull on the rope with one hand and, at the same time, cup the other hand over the bent back claw (*photo 19*). This protects the neck of the uterus as the foot comes forward and is an extremely important precaution against injury to the mother.

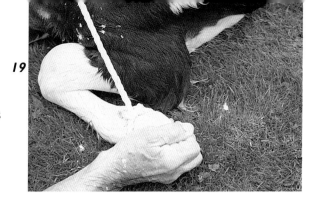

Pull the foot outside the vulva (*photo 20*) and when both legs are forward, rope both feet, this time above the fetlocks.

Get an assistant to pull on the legs of the foetus while you use one hand to cover the neck stump to prevent injury to the roof of the vagina. Keep the neck stump covered until it protrudes from the vulva. Once the calf's shoulders are out, the rest is easy (*photo 21*).

Exactly the same principles are adopted if both legs are back. The stump is pushed back as far as possible into the uterus, one foot at a time is roped around the pastern, and one hand is cupped over the foot as it is pulled forward through the cervix. If the foetus is not pushed back, correction is much more difficult and, in a heifer, well nigh impossible.

If the calf is alive, probably only a veterinary surgeon will succeed in delivering it alive. He will most likely use a calculated dose of spinal anaesthetic — usually 4½ to 5 cc of a 2 per cent solution. This is sufficient to stop the cow straining, but not enough to take away the power of her legs.

The veterinary surgeon will now wash and lubricate the protuding parts of the foetus, get the cow onto her feet if possible, and gently push the head and leg or the head back into the uterus. He will then bring forward the turned back leg in the manner described.

The spinal anaesthesia may make the animal groggy for an hour or two, but if the floor underneath is sanded or gritted and she is hobbled she will come to no harm.

HEAD DOWN INSIDE UTERUS AND LEG(S) BACK

Another simple malpresentation is where the top of the calf's head is presented to the cervix, with the nose and mouth tucked down into the body of the uterus. One or both of the forelegs may also be turned back (*photo 22*).

First of all, introduce the running loop of a

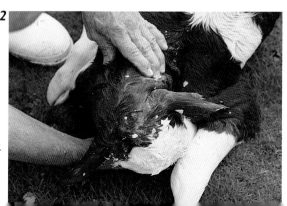

42

calving cord and pass it between the side of the head and the inside of the turned back limb (*photo 23*).

Now introduce the centre part of a looped cord (*photo 25*). Pass it over the top of the front of the head to underneath the lower jaw or chin.

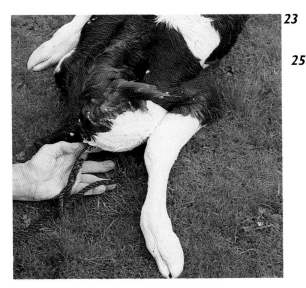

23

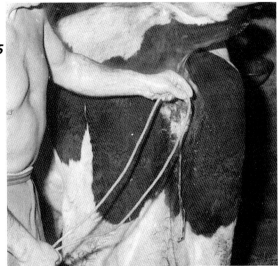

25

Apply the loop around the calf's coronet (i.e. between the pastern and the hoof) and bring the foot forward as instructed on page 42, cupping the hand over the foot as it comes through the cervix. Then, when the foot is forward, slip the loop up over the fetlock (*photo 24*).

Place the flat of the left hand on the top of the head of the foetus. As the animal finishes each strain (and never during the strain), push the head back and, at the same time, apply pressure on the ends of the cord (*photo 26*).

The calf's head will straighten easily. As it does so, slip the left hand underneath the calf's

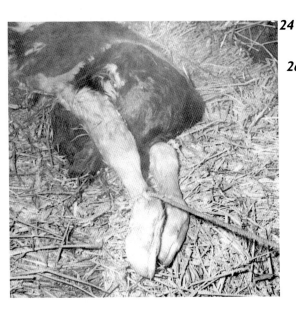

24

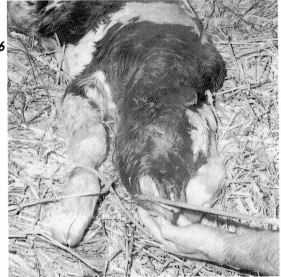

26

chin and help the chin over the pelvic brim
(*photo 27*). If pressure is applied during
straining, there is some danger that the incisor
teeth of the foetus may tear or rupture the
uterus.

When the positions of the head and legs
have both been corrected, the birth will
proceed normally.

28

27

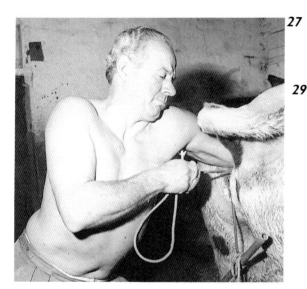

29

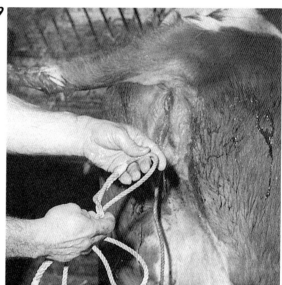

BREECH PRESENTATION

A breech is simply when the tail end of the
foetus is presented at or through the cervix or
vulva with the hind limbs of the foetus
extended downwards, upwards or straight
forwards into the uterus (*photo 28*).

This presentation is probably one of the
most common of all malpresentations. It is one
of the easiest to put right if the correct
technique is used, and one of the most difficult
to attempt if the correct procedure is
unknown. The way to deal with a breech
presentation is as follows.

First of all, make a running noose at one end
of a calving cord (*photo 29*).

Pass the noose between the turned back
hind legs of the calf as far as the nearest foot
that can be felt (*photo 30*). It may be
necessary to hook the hand into the hock joint
and pull it forward before a foot can be
reached.

30

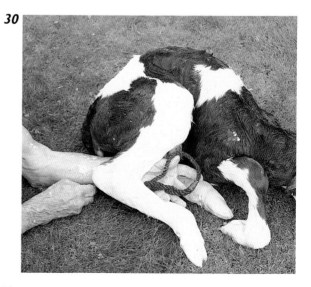

Slip the noose over the foot and up to just below the fetlock joint (*photo 31*). In many cases, in heifers particularly, it is better to secure the loop around the coronet, i.e. immediately above the hoof. Photos 18 & 19 illustrate the straightening of the leg.

The simultaneous pushing forward of the hock and maintenance of steady pressure on the foot, with slight added pressure between the mother's strains, will result in the hind foot shooting forward into the anterior vagina.

Correction of the second leg is much easier and in photo 33 you see the correct looping of the cord around the coronet.

31

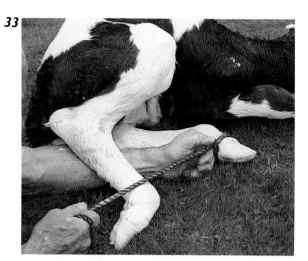

33

Place the flat of the hand underneath the hock of the foetus and push the hock upwards and forwards into the uterus, at the same time maintaining a steady pressure on the roped foot (*photo 32*). If a steady pressure is not maintained, the noose of the rope may slip up above the fetlock. When this happens, the foot does not turn backwards correctly and may tear the uterus.

Pushing the hock upwards and forwards, and simultaneous pulling on the foot, soon results in the straightening of the leg (*photo 34*).

It is a good idea during the straightening, particularly in a heifer, to pass the hand from the hock to the foot occasionally to make certain, first of all, that the noose is in the

32

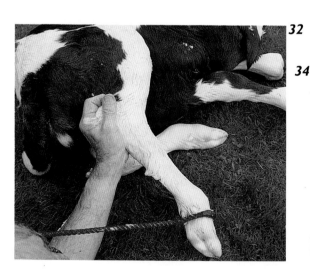

34

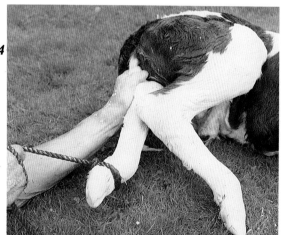

correct position and, secondly, that the front of the foot is not cutting into the wall of the uterus, as it sometimes does just below the cervix.

During the pushing back of the hock, steady pressure should be exerted on the rope, but **never excess** pressure. If excess pressure is applied, then obviously there is constant danger of the front of the foot tearing the uterus.

After both legs are straight and forward outside the vulva, move the loops up over the fetlocks before starting to assist (*photo 35*).

Presentation', page 29.

Just occasionally, and particularly with the younger and less experienced veterinary surgeon, spinal anaesthesia may be required before the hocks can be brought forward sufficiently to allow the flat of the hand to get underneath for the upward and forward propulsion. The disadvantage of spinal anaesthesia (*photo 37*), however, is that it interferes with the natural dilation of the cervix.

Less frequently, the joints of the hind legs may be stiffened or ankylosed, and it may not

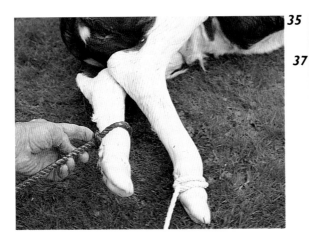

35

37

Later, when approaching the crisis stage (i.e. when the tail appears outside the vulva), the loops can be fixed above the hocks to give the assistant better leverage. Remember, once the hind legs of the 'breech' are straightened, you are dealing with a simple posterior presentation (*photo 36*) and you should proceed exactly as instructed in 'Posterior

be possible to either bend the fetlock or move the hock joint. In such cases embryotomy wire has to be introduced around the top of one or both hind legs and the entire stiffened limb or limbs removed (*photo 38*). Usually the removal of only one allows the delivery of the foetus with the remaining ankylosed hind leg still projecting forward.

36

38

46

7
How to Resuscitate a Calf

Many calves' lives are lost through a lack of simply knowing how to induce the all-important first gasp of breath.

Such loss of life occurs chiefly when the calf is born backwards, due to the fact that the navel cord ruptures while the head and chest of the calf are still inside the cow. Rupture of the cord usually occurs as soon as the entire hind end of the calf is outside the cow's vulva.

In practically every case when the calf appears to be dead, the heart is still beating. If you succeed in inducing a single gasp of air into the lungs, the calf can often be saved.

There are four simple methods of producing that breathing reflex.

STRAW IN THE NOSE

Insert a fairly rigid piece of straw into the calf's nostril and push it up as far as you like (*photo 1*). Continue to move the straw up and down inside the nostril for about five or six seconds. In many cases, the calf will shake its head and start to breathe.

THE KISS OF LIFE

If the straw fails to produce a reflex, open the calf's mouth, hold the tongue forward on the floor of the mouth with one hand, and blow down the calf's throat (*photo 2*).

In this, all you are doing is blowing carbon dioxide into the calf's respiratory system — carbon dioxide is a well known and very valuable respiratory stimulant. Continue blowing for at least a minute.

MANUAL PRESSURE ON THE CHEST

The calf should be placed on its brisket, with its fore legs stretched out in front and its head resting on them. Intermittent pressure should be applied with the palms of the hands over the back of the chest cavity. The best way to judge the correct position is to try to have the back of the palms over the diaphragm (*photo 3*).

When using this form of artificial respiration, it is a good idea to have an assistant administering the kiss of life once very 20 seconds.

This is of tremendous value to a calf and can be continued for five or ten minutes.

COLD WATER

Get an assistant to hold the calf upside down, or suspend it by means of a rope tied over a beam, and throw a bucket of cold water over the chest and head regions (*photo 4*). This may seem a bit drastic, but I have found it successful when the straw and carbon dioxide have failed.

If the calf is heavy it can be suspended over a fence bar while the mucus is cleared from the mouth and nostrils (*photo 5*).

DRUGS

There are several respiratory stimulants which can be placed on the back of the calf's tongue. Your veterinary surgeon will be pleased to prescribe an emergency supply. Personally I have found these of limited value. Nonetheless they can occasionally provide a spectacular result.

8
The Afterbirth

The passing of the afterbirth in the cow is the third stage in natural normal birth. When the afterbirth is retained, therefore, there is always a reason (*photos 1 & 2*) and the animal will almost certainly lose condition.

Causes of afterbirth retention may be:

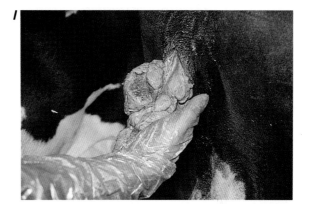

- Contagious, leptospiral, salmonella or fungal abortion
- Vibrio foetus infection
- Calcium deficiency
- Magnesium deficiency
- Blood poisoning, or septicaemia
- Inteference during calving
- Fatigue after calving

Causes of afterbirth retention

Contagious abortion
In contagious abortion, which is caused by the germ *Brucella abortus*, an inflammation of the lining of the uterus causes the afterbirth to adhere to the cotyledons or, as some people call them, 'the roses'. When the inflammation heals, it leaves fibrous or scar tissue between the cotyledons and the cleansing.

Contagious abortion afterbirths can hang for anything up to ten days, or even a fortnight. When they drop, or are removed, the parts that have been attached to the cotyledons usually have a characteristic yellow colour (*photo 1*).

Cows with retained afterbirths due to *Brucella abortus* have not necessarily calved before their time. **Contagious abortion**

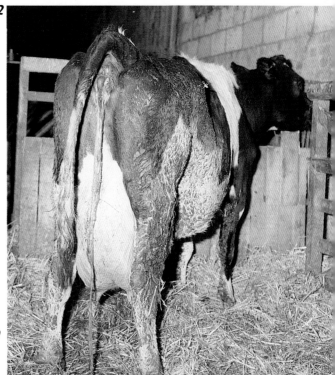

49

does not always cause abortion; the only sign that a cow is infected may be the holding of the afterbirth. Hence, if several cows in the herd are hanging on to their cleansings, blood samples should be taken immediately.

Fortunately in Britain, contagious abortion has been virtually eliminated. Nonetheless it is a legal obligation to have all suspect cases blood-tested for both Brucella and Leptospira (*photo 3*). At the same time samples of the uterine discharge should also be sent to a laboratory to check for both Leptospira and salmonella.

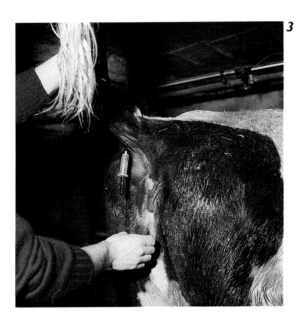

3

Leptospiral, salmonella and fungal abortions
Since the virtual eradication of brucellosis, leptospirosis is now recognised as one of the main causes of abortion in British cattle. In pregnant animals abortion is most likely to occur two to three months after infection; and cows in the later stages of pregnancy are the most vulnerable.

With salmonella there are usually, but not always, severe general symptoms like haemorrhagic (i.e. bloody) diarrhoea and a high fever.

Fungal or mycotic abortion in cattle can occur after eating mouldy hay, straw or poor quality silage containing infective spores (see pages 72-73).

Vibrio foetus infection
In vibrio foetus infection, the most serious of the bovine venereal diseases, the worst type of retentions occur when the calf is carried to full term. They are often impossible to remove in under ten days, the cotyledons seem to die inside the uterus, and invariably you are left with a really sick cow, which stands little chance of breeding again if the veterinary surgeon is not consulted.

When vibrio cases abort, the afterbirths are also retained and the cotyledons die. The dead cotyledons subsequently come away with the afterbirth and an evil smell may persist for several weeks. The salvation of such cases lies in the daily injection, for at least a week, of massive doses of antibiotics.

The majority of vibrio cases, of course, don't hold to the bull or break at 10 to 12 weeks. Nonetheless, a great many infected cows — certainly more than is generally realised — carry their calves to full term and hold their cleansings.

Calcium deficiency and magnesium deficiency
Nerve endings in the uterine muscle are surrounded by a solution containing both calcium and magnesium (*see diagram*). Even a mild shortage of either mineral prevents the uterus from functioning normally, thus causing retention of the afterbirth.

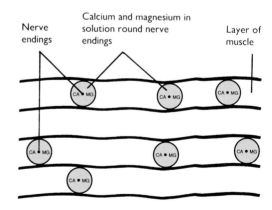

Nerve endings | Calcium and magnesium in solution round nerve endings | Layer of muscle

Blood poisoning
In summer mastitis or in gangrenous mastitis, for example (*photo 4*), germs in the blood excrete waste products called toxins. These toxins poison the blood and damage the

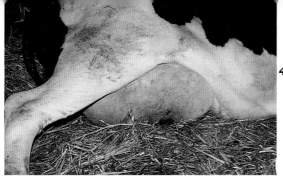

uterine muscle fibre, preventing normal uterine contraction.

When toxins build up in the blood system the condition is called toxaemia, and this toxaemia is very likely to kill the calf. I'm sure many readers have seen acute summer mastitis or acute gangrenous mastitis leading to a dead calf and a retained afterbirth.

Interference during calving
If the normal uterine contractions are disturbed this may lead to retention. Also the damage caused by interference often makes the vulva, vagina and cervix of the mother so tender and sore that she will make no attempt to void the afterbirth.

Fatigue after calving
Occasionally seen after a protracted labour, fatigue after calving is most often manifest when there are twins. Certainly the combined weight of two calves, often well over 50 kg, is more than enough to fatigue any muscle. This is the simple reason why a retained afterbirth is one of the hazards of twin calving (*photo 5*).

4   What should be done

The most important thing is to safeguard the animal's future breeding potential.

In order to do this, interference should be restricted to an absolute minimum. There is little doubt that a hand moving about inside a uterus for any longer than, say, 15 minutes at the outside is practically certain to set up a traumatic inflammation and, once this happens, infertility is the likely result.

My advice is as follows.

Where there is only an occasional case in the herd of a retained afterbirth, clip off the hanging portion (*photo 6*) and wash the hindquarters and udder with hot water, soap and antiseptic.

Disregard the portion of afterbirth left inside unless or until the cow goes off her milk or off her food. Then, of course, it is a job for your veterinary surgeon. He will remove the afterbirth or, if the cleansing is not quite ready, he will inject the cow intramuscularly with a hormone and may or may not prescribe a course of antibiotics, depending on how bad the case is.

Where cases of retention are occurring frequently in the herd, consult your veterinary surgeon immediately. He should be able very quickly to pinpoint the predisposing cause and prescribe a cure.

The thing to remember is that an occasional retained afterbirth is inevitable, whereas frequent cases are abnormal and indicate a herd problem.

5

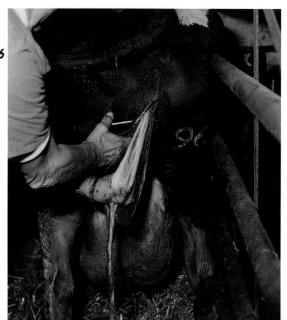

6

9
Spinal Anaesthesia
(Epidural Anaesthesia)

The use of spinal or epidural anaesthesia is definitely a job for the veterinary surgeon. Nonetheless, a thorough knowledge of why and how it is used will not only be of value to the veterinary student, but will help the lay reader to more fully understand this aspect of a veterinary surgeon's work.

Local anaesthetic is injected into a space, called the epidural space, which lies at the base of the spinal column at the top of the cow's tail. The equipment required is illustrated in photo 1.

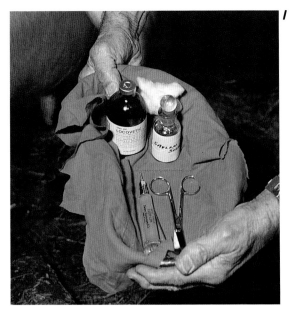

1

Uses
Spinal anaesthesia is used in the following cases.

- Caesarean section operation
- Prolonged interference
- Prolapsed uterus
- Any local surgery at the back end of the cow

In the caesarean section, which is done with the cow standing up, spinal anaesthesia prevents the cow continually straining and relaxes the uterus when the surgeon comes to handle it. Tranquillisers like xylazine can be used in caesarean sections instead of spinal anaesthesia.

In difficult cases, where prolonged manipulation or extensive embryotomy is required, the spinal anaesthesia not only makes the job infinitely easier for the operator, but goes a long way towards minimising the risk of shock to the patient.

In a prolapsed uterus (see 'The Problem of Prolapse', page 61), spinal anaesthesia can be vital because it stops the cow straining and makes the the prolapse return very much easier.

In local rear-end surgery it removes all pain.

Technique
The site generally recommended is the space between the first and second vertebrae of the tail. The site I like best is the space between the fixed end of the spine (the sacrum) and the

movable tail. The best way to find the site is to move the entire tail up and down with one hand, at the same time feeling with the thumb of the other hand for the space at the first point of movement (*photo 2*).

4

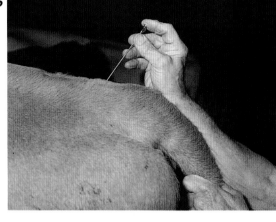

Now take an epidural needle, i.e. a needle specially made for the purpose, which has a central cannula to avoid needle blockage. Again, this needle must be absolutely sterile (*photo 5*).

2

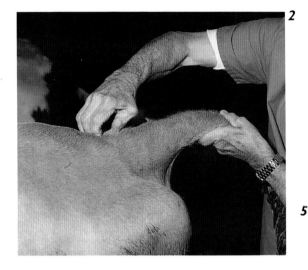

5

Having established the site, clip all the hair over the area (*photo 3*).

Now disinfect the site thoroughly with a powerful skin antiseptic (*photo 4*).

This is tremendously important since any infection taken into the epidural space (i.e. the space at the end of the spinal cord into which the injection has to be made) can cost the life of the cow.

3

Keeping the thumb on the vertebra immediately above the space selected, insert the needle, at an angle of approximately 45 degrees, boldly downwards (*photo 6*). The point will come to rest on the floor of the spinal canal.

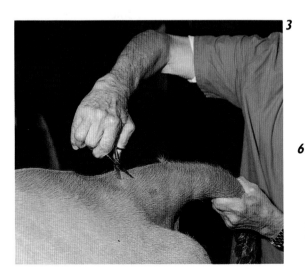

6

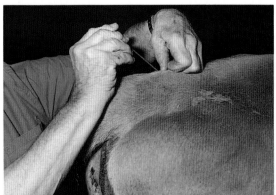

Fill the syringe with the anaesthetic. The
average dose required to stop a cow straining
is around 4½ cc of a two per cent solution of
a sterile local anaesthetic (*photo 7*).

You will know if and when you are in the
correct spot when you remove the cannula
from the epidural needle — you should hear
the hiss of air being sucked into the space in
the spinal column.

You will be able to confirm further the
correctness of your site by the ease with
which the anaesthetic can be injected
(*photo 8*). If there is any resistance to the
plunger at all, then the needle should be
withdrawn and another site tried.

The anaesthetic acts almost immediately and
within seconds you should be able to detect a
loss of power in the cow's tail. If a dose larger
than 4 to 4½ cc is given, there is great danger
of the cow losing the power of her hind legs.
This is undesirable, especially in a caesarean
section operation, where the whole job is
much easier if the patient remains standing.

7

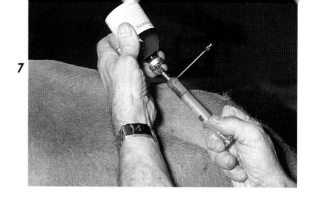

8

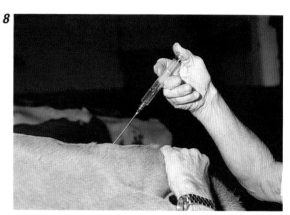

10
Torsion or Twist of the Uterus

Occasionally, during pregnancy, the uterus
containing the calf becomes rotated on its own axis.
This causes a torsion or twist of the uterine body
which is usually manifest in the anterior vagina. The
exact cause of uterine torsion is obscure, but it
may be due to a fall or to the excessively vigorous
movements of a big calf.

The torsion may only be partial or it may be
complete, depending on the degree of
displacement of the calf. The diagram shows a
complete torsion.

Correction of a twisted uterus must be carried

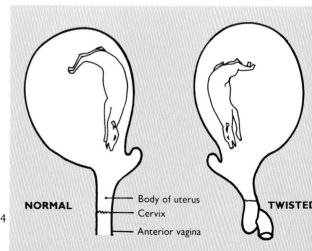

NORMAL   Body of uterus   TWISTED
                Cervix
                Anterior vagina

out by a veterinary surgeon. However, since stockmen will be required to help, a knowledge of what the vet is trying to do should increase the farmer's interest and enthusiasm.

Symptoms

The symptoms are not easy to detect. The cow goes to full term, or perhaps a few days over. She fills her udder and relaxes in the pelvic bones. She then starts showing signs of preliminary labour — mild bouts of labour pain, tail cocking and slight straining (*photo 1*).

These symptoms persist and continue without any further progress being made. After 24 to 48 hours, the cow becomes dejected and off her feed. Her extremities become ice-cold.

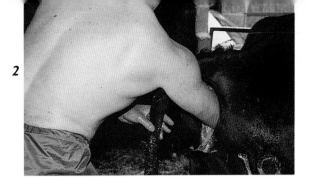

How to deal with it

First of all, the cow is roped for casting by Reuff's method: with a fixed noose around the base of the neck and in front of the shoulder (in horned cattle this can be a running noose around the base of the horns); a half-hitch around the cow just behind the elbows; and a second half-hitch around the body of the cow just in front of the udder (*photo 3*).

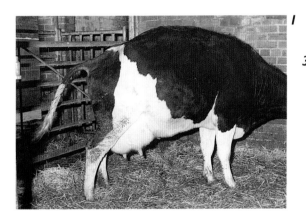

What to do

The diagnosis and treatment of uterine torsion should always be left to the veterinary surgeon, but it is very important that the farmer spots the condition as early as possible. Obviously the quicker the veterinary surgeon is called the easier the job will be, and the greater the chance of delivering a live calf.

Consequently, in all cases of abnormally protracted first-stage labour, especially in a cow, a vaginal examination should be made (*photo 2*). When the hand reaches the anterior vagina a queer tight-banded obstruction will be felt. Further introduction of the hand over the band will reveal the corkscrew torsion. In most cases the twist will be clockwise in direction.

If the anterior vagina feels in any way abnormal, send for your veterinary surgeon at once.

The cow is cast by pulling tightly on the free end of the rope from directly behind the rear end (*photo 4*). The tightening rope acts by exerting pressure on the blood vessels and nerves which supply the legs.

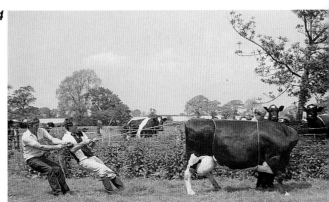

Three-man pressure may be needed, but eventually and certainly the cow will sink down. During the pulling it is essential to have the cow's head held firmly by a halter or by the nose. In other words, for this job plenty of help is needed — the more the better (*photo 5*).

The rolling is continued until the cow is on her back (*photo 8*). At this stage the veterinary surgeon will strip off, scrub up and insert a lubricated hand and arm into the vagina to assess whether the clockwise torsion is correcting itself.

Considerable mauling and continual turning

5

8

The cow is now rolled in the direction of the twist — in the case illustrated, in a clockwise direction (*photos 6 & 7*).

over may be required before the twist sorts itself out. In some cases, rolling first one way, then the other may have to be tried and, occasionally, the cow may have to be rolled over several times in the one direction. Eventually, however, in 99 cases out of 100, the torsion will straighten.

With a partial torsion and the cow standing, it is often possible to grasp a leg of the calf and rock the calf from side to side with sufficient strength to correct the torsion.

The cow can then be freed and either calved or left to get on with the job by herself, depending on the degree of dilation of the cervix. Most cases have to be left at least for a time after correction (*photo 9*).

If the twist is irreducible, then a caesarean section operation is necessary (see next chapter).

6

7

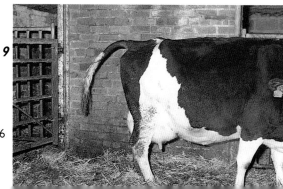

9

11
Caesarean Section

The surgical operation of caesarean section in a cow is very much a matter for the veterinary profession. However, I think it is well worth recording a description here, not only for the benefit of veterinary students, but in order that farming readers will understand and appreciate the technique and will, in comparatively rare cases when the need arises, be able to co-operate more fully with their veterinary surgeon.

The important thing to remember is that, provided the general principles so far outlined in this book are rigidly adhered to, the need for caesarean section will arise only occasionally.

REASONS FOR THE OPERATION

The caesarean operation is necessary in the following instances.

- An irreducible torsion of the uterus.
- An oversized foetus. This is not common, but when it occurs, caesarean is the only answer.
- Certain cases of schistosomus reflexus, especially in first calvers, where all four feet of the monstrous calf are presented (see photo on page 77).

Since the vast majority of caesarean section operations are performed under farm conditions, this colour sequence on pages 57-61 illustrates the routine technique, now adopted by my partners, using an intravenous tranquilliser instead of spinal anaesthesia.

Two veterinary surgeons working in unison greatly reduce the time factor besides allaying any doubt in the farmer's mind regarding the necessity of surgery.

One word of advice on bovine obstetrics. Perhaps the best way of all to minimise calving difficulties in any breed is to refrain from any cross-breeding, especially in heifers.

A wise precaution: Get the opinion of two veterinary surgeons before resorting to a caesarean section, as normal birth is always preferable if possible. In any case the operation is best performed by two vets.

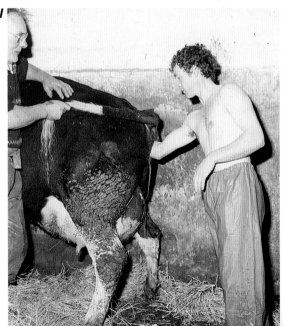

57

Sedate the cow with a tranquilliser injected intravenously using the tail vein. Give a small dose only since the cow must remain on her legs.

Inject local anaesthetic to one side of the entire length of the proposed incision — under the skin, into the muscle and into the peritoneum.

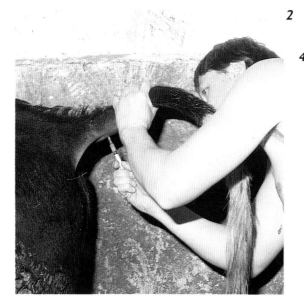

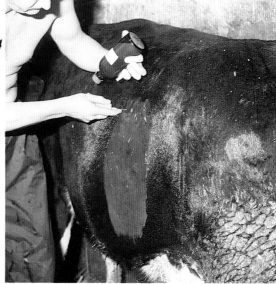

Shave the entire operation area, scrub with soap and water and non-irritant antiseptic and then sterilise the skin surface with an iodine preparation or Merthiolate.

Using a sharp sterile scalpel blade, make a 32 to 37 cm incision through the skin.

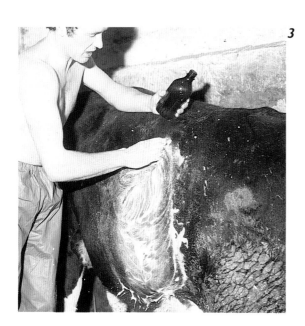

Now using sharp sterile scissors, cut through the muscles.

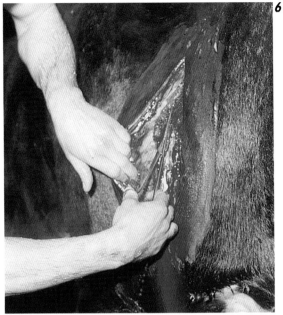

6

Attach a sterile scalpel blade by nylon to the wrist and then insert a well-lubricated hand and arm and identify the gravid uterine horn. Make a careful incision as far forward as possible and withdraw the first hind foot of the calf.

8

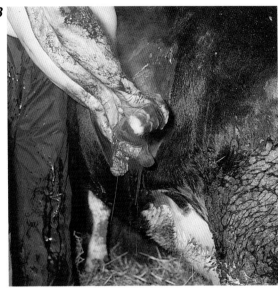

Still using the same scissors, cut through the peritoneum.

7

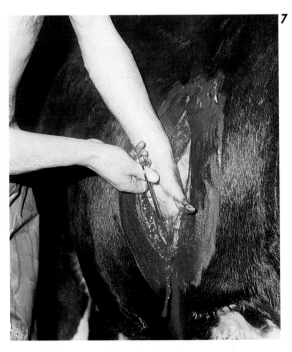

Find and withdraw the second hind foot.

9

Now with the farmer's help, withdraw the calf as quickly as possible. The size of the hips in this calf justified the caesarean section, especially since the mother was a heifer.

10

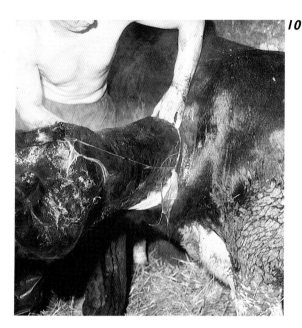

Identify the uterus and withdraw it from the incision.

12

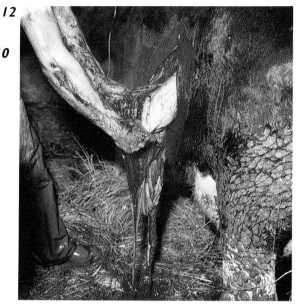

A live calf! It makes the emergency surgery really well worthwhile.

11

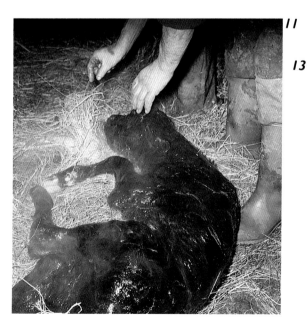

Carefully suture the wound with extra-strong catgut by stitching the peritoneal surfaces of the uterus into apposition (the Czerny-Lembert suture). This way adhesion and healing are very rapid.

13

Close the muscle and peritoneum with a continuous catgut suture.

Finally, close the skin wound with a series of mattress sutures using strong nylon.

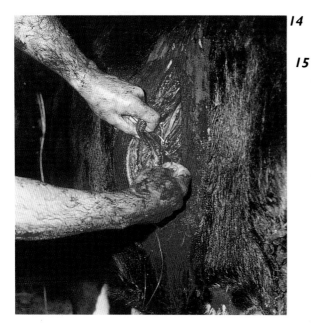

14

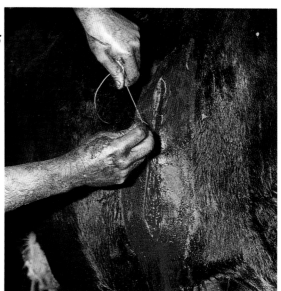

15

12
The Problem of Prolapse

PROLAPSED CERVIX AND VAGINA

A prolapsed cervix usually occurs when the cow is heavy in calf. In the early stages, the cow shows only a portion of the cervix and this mainly when lying down. Later, the prolapse may become an angry-looking red mass, even though when the cow stands up the prolapse sometimes disappears (*photo 1*).

When the calf gets close to full term — or if, as often happens, the urinary bladder

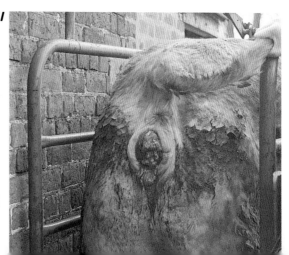

1

becomes incorporated inside the prolapse — then the prolapse stays out and gets bigger and angrier-looking the longer it is left without attention. What happens is that the ligaments and muscles which normally hold the vagina and bladder in position become stretched or torn due to the weight of the extended uterus (*photo 2*).

2

What to do
The main thing is not to panic — a cow which 'shows her reed' before calving very rarely prolapses the cervix, vagina or uterus after calving. **Despite this fact, it is unwise to keep the cow for subsequent breeding, since the trouble will reappear earlier in the following pregnancy and will be correspondingly more difficult to deal with.**

How to deal with it
This is definitely a job for the veterinary surgeon. He will probably use spinal anaesthesia, then he will wash and dry the prolapse, dress it with a suitable antibiotic in oil or sulpha dressing, and replace and suture it in position.

He will most likely use a deep mattress suture (or stitch) — at least, this is what I prefer. Using strong silk or nylon, he will pass the needle as deeply as possible through both sides at the top of the vulva (taking a good 'bite' of the skin alongside).

Then he will bring the needle to the bottom of the vulva and, taking great care not to include the urethral orifice (i.e. the hole leading into the bladder), will pass the needle right back through both sides, this time in the opposite direction and again as deeply as possible. He will then tie a reef knot on the side of the original entry of the needle (*photo 3*).

This stitch will last for at least 14 days but, after that time, it may have to be renewed if the patient still hasn't calved.

3

PROLAPSED UTERUS

A prolapsed uterus is not common. In fact, many farmers go through life without a single case and those who do see one often regard it as a once-in-a-lifetime experience (*photo 4*).

4

62

However, there are exceptions and a number of cases may occur on the same farm. The explanation of this may lie in the chief predisposing cause in cows: calcium deficiency. A lack of calcium causes the tone of the uterine neck (or cervical) muscles to be lost and this allows the whole uterus to come away.

What appears to happen in many cases is that the milk fever associated with the calcium deficiency makes the cow lie flat on her side, the rumen blows up and, when the blown, consitpated, calcium-deficient cow strains to pass dung, the abdominal pressure pushes the uterus through the relaxed cervix.

Obviously, the incidence of calcium deficiency will vary from area to area and from herd to herd.

Does this hold good with heifers? The answer must be no. I have yet to see a uterine prolapse in a first-calver where the patient has shown any signs of milk fever.

Nonetheless, I have found it wise to inject calcium into heifers as well as cows before attempting the prolapse return.

The cause in heifers appears to be a simple turning inside out of the tip of the pregnant uterine horn. During the final expulsion of the calf, the tip of the horn folds inwards on itself, rather like the toe of a sock being turned inside out. Once this happens, the continuing uterine contractions increase the eversion and eventually push the lot out.

The predisposing cause in first-calvers is often excessive manual pulling at calving. By exerting a continual high-pressure pull, the uterus is not allowed to relax and fall back into its normal position between contractions. Obviously, this is likely to cause the tip of the horn to turn inward on itself.

How to recognise

If there is a huge red mass, dotted over with large lumps of cotyledons, protruding from the vulva after calving, you have a prolapsed uterus. Often the cleansing is still attached to the cotyledons (or 'roses' as they are sometimes called).

The sight of this gives one a lasting knowledge of how the afterbirth is attached to the uterus (*photo 5*).

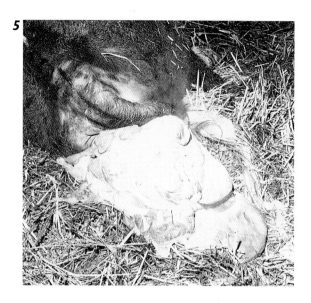

5

What to do

First-aid measures are simple.

Phone an urgent call through to your veterinary surgeon (these cases get priority from us).

If the cow is in a yard or cubicles with others, prompt action is urgent. Get a man to stand protectively over the prolapse while the other cows are turned out. I've seen several uteri ruptured by the trampling feet of cows (*photo 6*).

6

Then all that is necessary is to keep the prolapse as clean and as warm as possible. A clean sheet and blanket are quite adequate, but hot water bottles and hot towels are better (*photo 7*).

7

Don't, whatever you do, attempt to push the prolapse back, because you'll kill the cow (or yourself!) for sure.

If you must exercise your genius, and you have a flutter valve and some calcium handy, inject one bottle of calcium underneath the skin, high up and one hand's-breadth behind the shoulder (*photo 8*). Rub it well away, then sit back and wait for your veterinary surgeon.

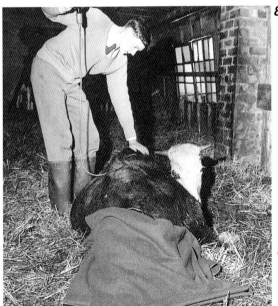

8

How to deal with it

Uterine prolapse return is one job where everything is needed: scientific knowledge, professional skill, experience, improvisation and a very considerable amount of physical strength. The use of spinal anaesthesia, antibiotics, absolute cleanliness, special drugs to contract and protect the uterus, non-irritant lubricants, and a technique devised by trial and error throughout the years, enables the veterinary surgeon to save the vast majority of uterine prolapse cases.

The technique which I have found best — and this information is for the veterinary reader particularly — is as follows.

Check and lay out all the tools for the job (*photo 9*), and check on coverage and warmth of the uterus.

9

Give a calcium injection intravenously (*photo 10*), and subcutaneously also if the owner hasn't already done this.

Inject approximately 5 cc of a two per cent solution of local anaesthetic into the epidural space. This will stop the animal straining against the return of the prolapse (*photo 11*).

Inject a dose of muscle relaxant intravenously or intramuscularly. This will particularly relax the muscles around the vulva and vagina. An intramuscular injection of an extract of the pituitary gland may be given to contract and reduce the prolapse size.

10

12

11 abdominal contents forward, leaving more room for the prolapse return and making the return infinitely easier.

Insert a pastry board or tray underneath the prolapse, with an assistant holding the ends of the board or tray on either side (*photo 13*). This is an excellent hint, because it means that the operator does not have to hold the weight of the uterus while he is pushing it back.

13

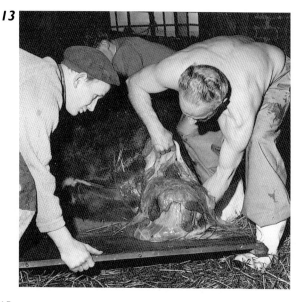

After five minutes, roll the animal's back end onto a sleeper or beam (*photo 12*). If this is not practicable and if the cow is in a cowshed, pull her head end round into the grip and roll her hind end over on to the bed.

Either of these precautions is essential for two reasons. Firstly, it allows the spinal anaesthetic to spread and act more completely and, secondly, it throws the weight of the

65

Another very successful prolapse replacement techique is to sit the patient on her belly and chest, pull both hind legs straight out behind her and tie them with a strong piece of rope (*photo 14*).

Wash the prolapse thoroughly (*photo 16*), and dress with antibiotic sulpha powder and lubricant. Again, the best lubricant is soap flakes.

Return the prolapse slowly and carefully,

14

16

The pastry board to support the prolapse can then be rested on the cow's hocks (*photo 15*).

Using this technique it is possible to effect the replacement without spinal anaesthesia since straining is reduced to a minimum.

using the flat of both hands and working continuously close to the vulva (*photo 17*). If the cow stands up during the process the job is much easier, provided the assistants continue to take the weight of the prolapse on the board.

15

17

When the prolapse is back in the abdominal cavity, it is vitally important to make sure that the horns of the uterus are invaginated completely (*photo 18*). If this is not done, the patient will strain the prolapse out again. Once the prolapse is returned correctly, it should never come out again.

18

Despite this, it is always wise to insert a deep mattress suture (*photo 19*) as illustrated in 'Caesarean Section', page 57.

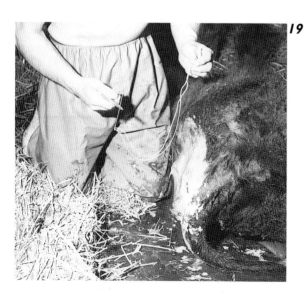

19

The two main enemies of the veterinary surgeon are shock and bowel prolapse.

The shock syndrome is most often seen in older cows, or where the uterus has been out for a long time. Usually such patients offer little or no resistance to the prolapse return, but only rarely do they survive. They refuse to eat or drink after the operation, their extremities are ice-cold, and often they breathe heavily or grunt ominously. I think such cases are best slaughtered, though occasionally I have pulled the odd one 'back from the grave' with a blood transfusion.

When the bowel comes out inside the uterus, then the replacement really is a dreadful, heartbreaking, and often impossible job, though few veterinary surgeons will give up without a long and exhausting try.

Occasionally gangrene sets in, but even then something can be done. The gangrenous uterus can be amputated with reasonable hope of success.

I always find that if a cow eats and/or drinks after a prolapse return or amputation, she will live and do well. If she won't show any interest, then she rarely survives.

A three-day course of intramuscular antibiotic completes the treatment and ensures the maximum chance of the cow subsequently holding to the bull. To my mind, this is very important.

The prolapse may, and often does, occur outside in total darkness where there is only a torch to assist the veterinary surgeon and very likely only one assistant.

Using the leg-back technique the photo sequence on pages 68-69 demonstrates how to deal successfully with such an emergency.

Two final important practical points.
Never scrap a cow after a uterine prolapse — rarely if ever will the cow evert her uterus twice. This may well be due to adhesions being set up inside after the mauling of the first return, adhesions which anchor the uterus in its correct place.

Never — and I repeat never — resort to excess pulling when calving a heifer. Remember, one man and patience are all that are needed, assisting only when necessary, and then only when the animal strains.

The prolapse.

The second hind leg of the cow is pulled back.

The cow is injected intravenously and subcutaneously with calcium, sat on her brisket with the single assistant astride her back, and one hind leg is pulled back without the help of a rope. The prolapse is laid on a rubber sheet.

Sitting astride the pulled-back legs of the cow and with the prolapse in his lap, the veterinary surgeon removes the afterbirth.

Having washed the prolapse in warm water containing a non-irritant antiseptic, the veterinary surgeon lubricates it first of all by coating the surface with antibiotic in oil and secondly by using a copious quantity of soap flakes.

24

The successful replacement is now effected.

25

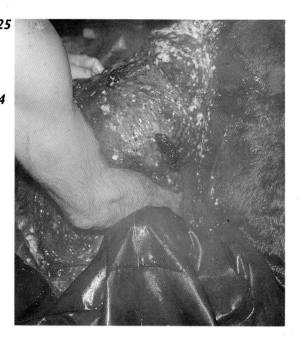

13
Hydrops Amnii (Uterine Dropsy)

Hydrops amnii or uterine dropsy is a condition where the foetal membranes surrounding the calf accumulate a massive quantity of oedematous or dropsical fluid (*photo 1*).

Symptoms
For a considerable time the cow looks as though she is carrying twins.

Towards the end of the gestation period the uterus and abdomen become so grossly distended that the cow has great difficulty in

1

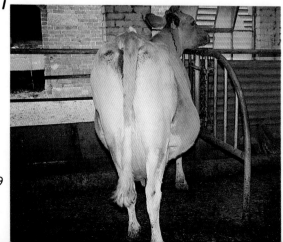

rising or walking about (*photo 2*). She may
start to grunt and go off her food. Her
extremities become ice-cold.

Treatment
If the condition is diagnosed early enough,
there are two alternatives: immediate
slaughter for salvage or the termination of the
pregnancy.

   The latter can be done by hormone
injections or by opening up the cervix and
rupturing the membranes.

   If the cow does go to full term, there is a
very grave danger of severe shock when she
loses the massive fluid content of the uterus.
Salvage slaughter at this later stage is rarely
economical because the oedema has often
extended into the carcase.

   Early diagnosis and prompt co-operation
with your veterinary surgeon are therefore
vital.

14
Infertility

The only sensible and economic way to control
infertility, particularly in the dairy herd, is to
get a veterinary surgeon to carry out a regular
routine pregnancy diagnosis. With non-
pregnant animals it will be possible to detect
and treat trouble in the early stages (*photo 1*).

The Cow's reproductive system
Before one can begin to understand the
problem of infertility, a knowledge of the
simple physiological function of the cow's
reproductive system is absolutely essential.

   First of all, the pituitary gland, a small gland
at the base of the brain, secretes a follicle-

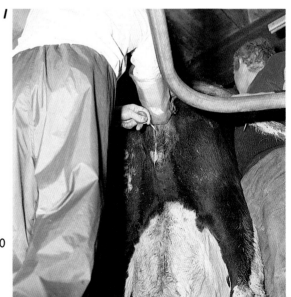

70

stimulating hormone. This stimulates the production of follicles in the ovaries, i.e. small cavities which contain the eggs or ova (*diagram 1*).

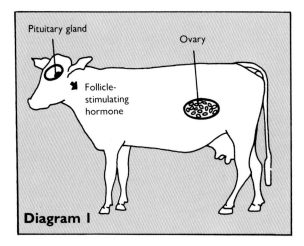

Diagram 1

Once every three weeks the ovaries themselves manufacture a hormone called oestrin. This produces the oestrus or heat period (*diagram 2*).

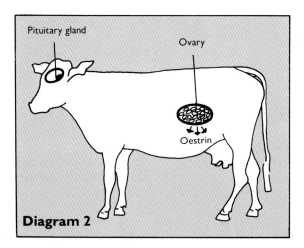

Diagram 2

Towards the end of the heat period the pituitary gland gets going again and this time secretes another hormone, the luteinising hormone (*diagram 3*). This is carried via the bloodstream to the ovaries where it causes the follicle or follicles to rupture, thereby releasing the egg or eggs which drop down into the

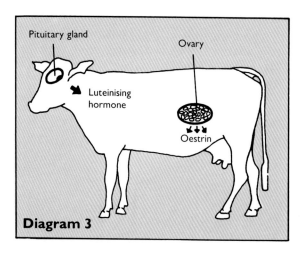

Diagram 3

oviducts or fallopian tubes. There they await fertilisation by the spermatozoa either from the bull or from A.I. (*diagram 4*).

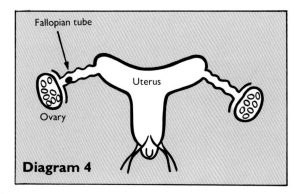

Diagram 4

If conception does take place, a small body called the corpus luteum develops in the ovary and manufactures yet another hormone, progesterone (*diagram 5*). Progesterone gets into the blood and, in its passage through the pituitary gland, it stops the production of the follicle-stimulating hormone.

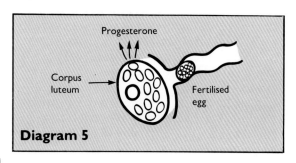

Diagram 5

This breaks the natural sexual cycle and the cow does not come into season again until the ovary sheds the corpus luteum, usually about the eighth or ninth day after calving (*diagram 6*).

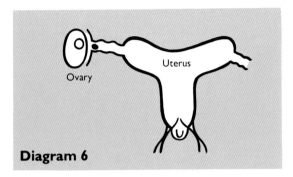

Diagram 6

Conception takes place in the upper part of the oviduct. Three or four days later the fertilised egg comes down into the uterus where it starts to grow into the foetus (*diagram 7*). In 30 to 40 days after fertilisation, the foetus becomes attached to the cotyledons which grow from the lining of the uterine wall.

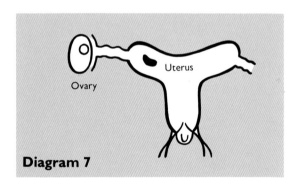

Diagram 7

Causes of infertility

Any factor which interferes with any stage of the normal sexual cycle can produce infertility, e.g. a retained afterbirth. The simple answer to all these problems is a professional consultation with your local veterinary surgeon as soon as possible.

Most infertility problems are due to what scientists describe broadly as non-infectious infertility, but I've often described them in language a great deal more expressive.

Anoestrus

One of the most common 'breaks' in the sexual cycle is the condition of anoestrus, where the cow fails to come into season. This is due merely to a low or negative output of follicle-stimulating hormone.

In the dairy cow the cause is usually a high milk yield on a comparatively low protein diet especially when the cow is still growing, which she does up to her third calf. Occasionally, it is due to a relative mineral deficiency — deficiency of phosphorus or of the trace elements cobalt, manganese and copper. An imbalance of molybdenum can produce the copper deficiency.

In the heifer it can be caused by malnutrition, mineral deficiency, parasitism, or cold or lack of shelter. In fact, any debilitating factor or disease, such as Johne's disease, can lead to anoestrus.

The same factors, especially a high milk yield on a low protein diet, can interfere with the production of luteinising hormone. When this happens the heat periods last for up to several days and the cows rarely hold to the bull.

One other very common cause of winter anoestrus and failure to hold to the bull is low blood sugar. This is often caused by feeding excess protein in relation to the carbohydrate intake.

Such a problem can be established by taking random blood samples from the herd — say six from newly calved cows, six from cows halfway through their lactation and six from cows nearing the end of the 365 days. A metabolic profile of such bloods will show clearly any blood sugar deficiency. The condition can be rectified by feeding 1kg (2lb) of maize meal per head per day throughout the entire winter.

Infection

Obviously an infection in the uterus or fallopian tubes will 'gum up' the works (*diagram 8*). Infection could be caused by contagious leptospira or salmonella or by one of the venereal diseases — trichomoniasis, vibriosis or coital exanthema ('bull burn'). Luckily the use of A.I. has largely reduced these venereal disease to a minimum.

One other infection is due to the fungus which causes mycotic abortion. This fungus is

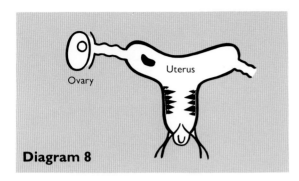

Diagram 8

found in mouldy hay (*photo 2*), straw or poor quality silage, so after a wet summer and a bad harvest don't feed mouldy fodder to your breeding stock. Keep only the best hay for the in-calvers and the milkers; mouldy hay can be used up on the store cattle. I can assure you mouldy fodder can be more dangerous to your herd than venereal disease, so be careful. This very simple hint can save a lot of money.

2

Cystic ovaries

But the biggest headache of all, and one that is becoming an increasing nightmare to progressive dairy farmers and veterinary surgeons alike, is the condition of cystic ovaries.

The cystic condition develops when ovulation does not take place, i.e. when either the egg doesn't develop correctly within the follicle or more commonly when the follicle fails to rupture when the egg is ripe. This is due to a dysfunction of the pituitary gland leading to a lack of luteinising hormone.

What causes cystic ovaries? Without a doubt the most common stress factor producing this ailment is a high milk yield coupled with a comparatively low plane of nutrition. This, I am certain, is the reason why we are seeing more and more cystic ovaries on dairy farms.

The trouble is that cows have been bred for high milk production and are expected to produce high yields either on grass or on grass products subsidised only by low protein cereals. Very often cysts first develop in the two or three months after the third lactation when the cow is reaching her peak production.

Just occasionally cysts can be hereditary. They can also be caused by a deficiency of the trace element manganese which gets locked up in the soil when too much nitrogen is used to boost young spring grasses. The deficiency persists in the silage and hay and builds up as the winter progresses and other stress factors such as cold and exposure take a hand.

But nearly always the cause is the persistent taking of gallons out of the pint pots, so my advice is to use commercial or hybrid cattle and be content with lower yields where the problem has become excessive.

Retained afterbirth (photo 3)
See 'The Afterbirth', page 49.

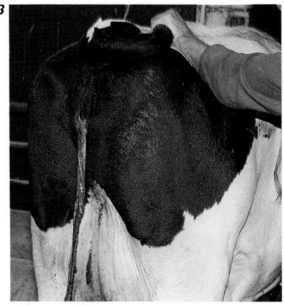

3

Poor heat detection
One common cause of infertility is failure to observe the animals in heat. This is especially the case with heifers and beef breeds.

Synchronisation of heat periods
There are hormonal products available which can synchronise the season. The two most popular are a drug called prostaglandin (marketed under the name of Estrumate) and a coil impregnated with progesterone (marketed under the name of PRID) which is also excellent for cysts.

With heifers prostaglandin is the most convenient. A dose of 2 cc is given intramuscularly on the first day of heat and twelve days after, and the heifer is inseminated three or four days after that whether or not symptoms of oestrous are present. Obviously if the heifer shows heat signs before the third day or after the insemination, say on the fifth or sixth day, then it is wise to serve her again.

Incidentally, a 2 cc dose of prostaglandin is invaluable in clearing up dirty uterine discharges after calving.

The progesterone coil is very useful in beef cows. The coil is inserted into the anterior vagina (*photo 4*) and left for twelve days. Then

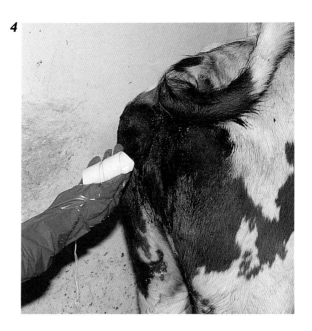

it is removed and the cow is inseminated or served two days later.

With both methods it is wise to have a fertile bull running with the animals concerned.

No doubt your veterinary surgeon will keep you updated on all new products as they come onto the market.

15
The Downer Cow

Frequently after a difficult or prolonged labour a cow is unable to rise. Such animals are referred to as 'downers'.

In the cow the paresis (paralysis) may be due to calcium, magnesium or phosphorus deficiency. Most of these cases can be treated successfully.

However, where excess traction has been used to withdraw a large foetus from a heifer or cow — excess traction by a gang of men, a pulley block, a tractor or by the use of the unscientific so-called calving aid (which has a similar action to the pulley block) (*photo 1*) — then the posterior paresis is most commonly

By tying the head loosely to both sides of the front of the apparatus there is no problem with feeding and watering the patient (*photo 3*).

See my other book *Cattle Ailments* for more detail.

due to damage to an obturator nerve supplying one of the hind legs, producing obturator paralysis. Such cases are extremely difficult to treat and many finish up in the knacker's yard. Certainly this has been my experience despite the use of the Bagshaw hoist and air beds.

During a visit to Barbados I saw an apparatus which will largely insure against such losses (*photo 2*). It was made by the veterinary surgeon Dr John Duckhouse who has kindly supplied me with the photographs.

Not only does the equipment enable one man to lift a downer cow off all four feet and transport her to a suitable box or yard, but — **and this is the important advantage** — the cow can be left in the apparatus with her feet on the ground for as long as it takes for her to regain the use of the damaged limb, without any danger of the further injury which inevitably follows when the animal is allowed to plunge about.

16
Foetal Malformations

A short leg and a club foot.

A dwarf calf.

The abnormal vulva of a freemartin, that is a heifer twin to a bull and most unlikely to breed.

An imperforate anus.

A successful operation on an imperforate anus — not always possible.

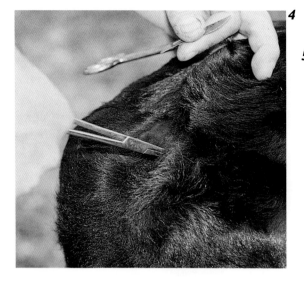

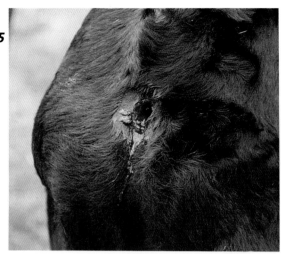

A schistosomus reflexus, often presented with the head and four abnormal limbs. In such cases a caesarean section is usually required. The foetal body is rudimentary. When the body is presented first, the foetus can usually be removed by embryotomy.

The typical hind leg movement of a spastic calf, a condition thought to be hereditary. Cases can be operated on successfully by cutting the tendons a few inches above the hock, but the operation is frowned upon by the breed societies.

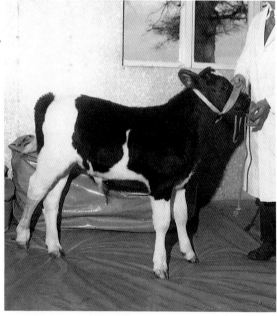

Care of the Calf

17
Introduction

The most important factors in caring for and preserving the life of a calf are as follows.

THE INTAKE OF COLOSTRUM

Colostrum is the first milk produced by the mother cow after calving. It is of vital importance: not only does it contain antibodies against most of the bacteria that the calf is likely to have to deal with during its most vulnerable period in life, but also, being much more concentrated than milk, it provides rapidly assimilable food containing vitamins A, D and E for the calf to store in its liver. It also contains a double dose of fat which acts as a laxative and enables the calf to pass its first dung called the meconium.

The calf is most capable of making full use of colostrum during the first four to six hours after birth so if it doesn't suckle its mother

during this period the colostrum should be milked off and given to the calf by stomach tube or by the ingenious method illustrated (*photo 1*). If the mother is ill or suffering from

1

mastitis, stored deep-frozen colostrum should be thawed out and given to the calf.

HYGIENE

In summer, calving at pasture probably provides the finest all-round conditions for the calf's survival. During the winter the cow should calve on its own in a calving box thickly bedded down with clean straw (*photo 2*).

When born the calf should be left to suckle the cow for at least three days or if this is not practical then at the very least 24 hours (*photo 3*). In the dairy herd the calf should then be transferred to a recently sterilised individual calf pen. With beef herds, of course, the calf will most likely be left with its mother.

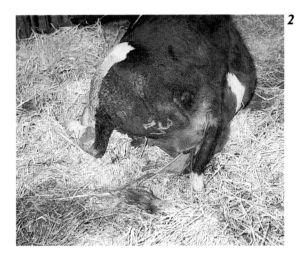 2

 3

18
Calf Housing

Perhaps the greatest contributory factor to calf losses is inadequate housing.

Environment is all-important to calves, particularly to young calves up to the age of three weeks, because during the first three weeks of life the heat-regulating centre in the calf's brain does not function properly. This means that the calf is not able to adapt itself to **sudden changes** of temperature.

Consequently, if it is exposed to such sudden changes, its entire resistance is lowered and it becomes susceptible to all the infectious diseases — scour, pneumonia, joint-ill, calf diptheria, etc.

The important thing, therefore, is to maintain the young calf at a constant temperature, or as near a constant temperature as possible.

This explains why calves born outside and reared outside are much less likely to contract disease — theirs is a constant environment and any change in temperature is rarely sudden.

The prevalent causes of rapid temperature changes in the average calf pen are draughts, lack of insulation and excessively high roofs which provide too much air space per calf.

Tumbledown multipurpose buildings, so often used for calf rearing, are literally death traps if not correctly adapted (*photo 1*).

abandoned double cowshed and provides ideal conditions for the hand-rearing of 40 calves. Photo 2 was taken during the summer when the false roofs over the calves had been set back and the ground draught was controlled by thick straw bedding at the front of the pens.

During my 53 years in cattle practice, I have repeatedly tried to obtain ideal conditions in the adaptation of existing buildings, always concentrating on the basic requirements of cleanliness, comfort, freedom from draughts, controlled access to fresh air and ample outlets for stale air. I have tried installing fans, air-conditioning and maximum and minimum thermometers without any spectacular success and never close to perfection.

It is only during the last five years that I have come across what can only be described as a near-perfect set-up with a proven virtually non-existent calf mortality.

Since this set-up incorporates all my ideals I take great pleasure in presenting it in the hope that it will inspire the reader to follow a similar routine and save themselves many thousands of pounds in calf losses. If building a special calf house, the same ideals should be aimed at.

The adaptation was of a previously

During the winter the false roofs are fully extended and the ground draught rigidly excluded by straw bales (*photo 3*).

Twice a year to dovetail with the calving index all pens are scrubbed out, disinfected and rested for one month. During the calving season the same principle is applied to the pens in rotation of tens (*photo 4*).

Warm dry bedding is essential and the main material used is ample straw (*photo 5*). Initially the straw was set on a layer of ashes, but with the cleaning routine established it was found that the ashes were not essential, especially since the original cowshed floor sloped outwardly.

Sawdust was tried and, although providing a comfortable warm dry bed, it was found to lack the cleanliness of the straw by adhering when damp to the calves (*photo 6*).

Lately the efficiency of shredded dust-proof

paper bedding, now widely used for horses, is being assessed. Undoubtedly the dustless paper provides an extremely comfortable bed, but the farmer has yet to satisfy himself as to its manural potential and/or disposal (*photo 7*).

Photos 8-10 demonstrate the preparation of each pen. Top quality hay is provided either

in filled racks or inside each pen (*photo 8*). Straw is still the bedding of choice.

Clean fresh water is provided in one bucket (*photo 9*).

Concentrates are supplied in the second bucket (*photo 10*). Milk or milk substitute is provided in separate sterilised stainless steel buckets.

The hinged gate provides easy access (*photo 11*).

The false roof comprises large mesh wire netting covered with thick wads of straw (*photo 12*).

For any premature or weakly calves an infra-red lamp is provided from a fixture plug directly over each pen. The lamp is adjusted to

8

11

9

12

10

approximately 45 cm above the standing calf's back (*photo 13*).

13

the eradication of pneumonia. In Canada and America this ideal is now being aimed at by the provision of individual plastic houses (*photos 16 & 17*); their success in Britain has yet to be proved.

14

Any calf showing the slightest sign of illness is immediately removed and isolated in a hospital box (*photo 14*) similar to that described in 'The Hospital Box', page 85.

Ventilation is well-nigh perfect with free movement of fresh air along the front and back of the pens but no through draught, and adequate height above the false roofs for the escape of any warm stale air (*photo 15*).

In wintertime the sliding doors at the top and bottom of the cowshed are kept closed, but adequate fresh air passes through the spaces at the door bases. When the temperature rises in the spring or summer at least one of the doors is kept open, preferably that away from the prevailing wind.

The outstanding feature of this adaptation is

15

16

17

19
The Hospital Box

I am convinced that **every stock farm in the world should have a hospital pen, or pens, in which the stockman can nurse a weak or sickly calf.** Daily on my rounds I see an appalling loss and wastage directly due to the lack of this simple facility.

A dry comfortable bed underneath, and the luxuriant warm radiance of an infra-red lamp above, can make all the difference between life and death to the calf which has been shocked by a protracted or difficult birth or by the toxins or dehydration of a debilitating illness.

THE IDEAL HOSPITAL PEN

The ideal layout is illustrated clearly in this series of photographs. The pen should be sited in a sheltered corner facing south and should be as far as possible from healthy calves or adult stock.

The accent is on a constant temperature, cleanliness, ease of access and freedom from draughts (*photo 1*).

The air inlets are baffled and close to the insulated low ceiling (*photo 2*).

The air outlet is sited to one side of the hospital pens with access to adequate top roof ventilation (*photo 3*).

2

3
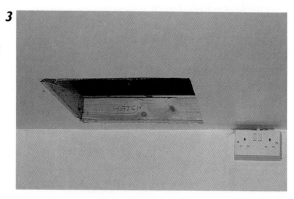

1

The floor, with a slight forward slope for drainage, is insulated by cardboard egg cartons covered by a thin layer of cement (*photo 4*).

A thick bed of straw completes the preparation. If available a generous bedding of sawdust would suffice though personally I prefer straw (*photo 7*).

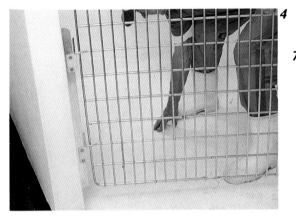

4

7

The back wall is rendered smooth to facilitate cleaning (*photo 5*)

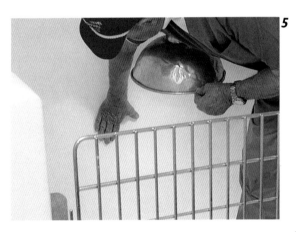

5

The sick calf is safely installed with the overhead adjustable infra-red lamp. If the calf is capable of standing the lamp should be at least 30 to 45 cm above its back (*photo 8*).

If the calf is dehydrated (as most of them are) the lamp should be switched off at least until the veterinary surgeon arrives to take over. He will probably supervise an intravenous drip and prescribe the correct course of subsequent treatment.

The bed is based with fine ashes or sawdust and the pen door is free from fixtures (*photo 6*).

6

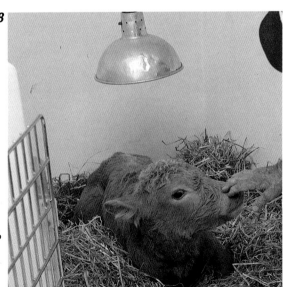

8

In such surroundings the calf will stand the maximum chance of recovery.

All feeding buckets should be sterilised and hay and water made available as recovery progresses.

THE IMPROVISED HOSPITAL PEN

For the improvised unit (*photo 9*), stick to the basic ideals recommended. For the walls, make use of a corner and straw bales. For the roof, either wire netting or wood slats covered over with sacks or straw. For the floor, loose ashes underneath the bedding.

Just two very important points: make absolutely certain that the improvised false roof fits tightly against the walls and on top of the bales to avoid draughts; and at all times guard against ground draught.

9

20
Calf Feeding

During recent years calf losses have assumed fantastic proportions. One local knacker-man told me that in less than a week he collected over a thousand dead calves, and this is just one single knacker's yard. If you consider such losses on a national scale, they become frightening — almost epidemic — and the position continues to worsen.

I am convinced that the losses are due almost entirely to a complete disregard for the simple physiology of the calf's digestive system. The disregard is not intentional, but is largely due to the pressures of modern farming. But one thing is certain: the losses will continue unless every stock farmer takes careful heed of everything that is written here.

NATURAL SUCKLING

When a calf is born its fourth stomach (the abomasum) is at least three times the size of its first stomach (or rumen) (*diagram 1*, page 88). The reason for this is simply that nature intends the calf to utilise its **fourth stomach** as its **main digestive organ** during the early part of its life.

When the calf starts sucking the cow, wagging its tail and perhaps bunting the udder but with its head held upwards, a groove called the oesophageal groove at the bottom of the oesophagus or food pipe forms into a closed tube (*diagram 2*, page 88).

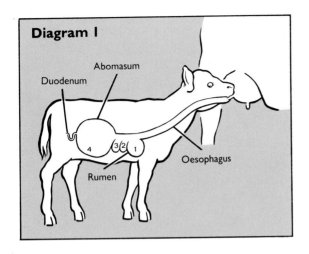

Diagram 1

Duodenum
Abomasum
Rumen
Oesophagus

4 | 3 | 2 | 1

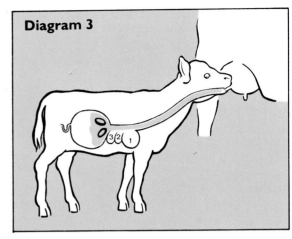

Diagram 3

3 | 2 | 1

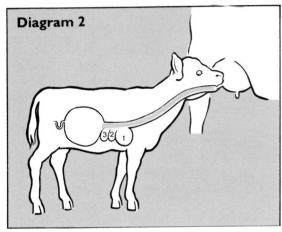

Diagram 2

3 | 2 | 1

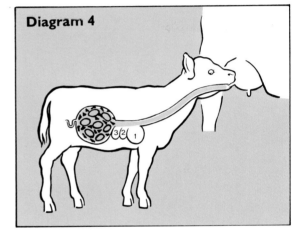

Diagram 4

3 | 2 | 1

The suckled milk then passes straight into the abomasum, bypassing the first, second and third stomachs. This is nature's design for calf digestion.

In natural suckling the calf feeds frequently — approximately every two hours — and each time it does a clot forms in the abomasum (*diagram 3*).

At the end of a day the abomasum contains a number of small clots, each one being acted on around the outside by the digestive juices (*diagram 4*). There is **absolutely no space left** for any other type of foodstuffs and this is **exactly how nature intended it.**

From two to three days old, the calf should be provided with ad lib hay, corn and fresh clean water. As the calf grows, it nibbles more and more of the hay and corn and instinctively drinks the correct amount of water. The hay, corn and water pass directly into the rumen or

first stomach. Gradually this stomach develops and increases in size so that when the calf is ready for weaning towards the end of the sixth week, the position is reversed — the rumen is at least three times the size of the abomasum and is ready and able to take on the main task of ruminant digestion.

BUCKET FEEDING

When a calf is being suckled out of a bucket with its head down, the tube formation at the bottom of the oesophagus is not so complete and a fair percentage of the suckled milk filters off into the rumen or first stomach where it ferments and is virtually wasted (*diagram 5*).

When the calf is being bucket-fed twice daily, as happens on most farms, only two clots are formed in the abomasum (*diagram 6*).

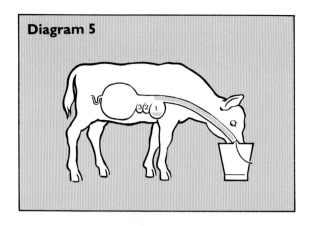

Diagram 5

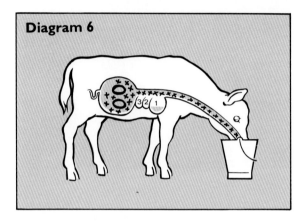

Diagram 6

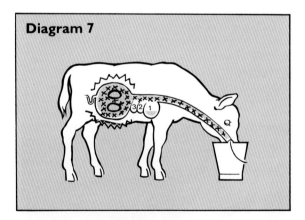

Diagram 7

These are obviously not as big as they should be and certainly they don't by any means completely fill the stomach cavity. What happens then? Because of the comparatively long periods between the feeds, the calf gets hungry; it nibbles more than its quota of hay and corn and perhaps drinks an excess of water.

Some of the fibrous foods, undigested, find their way through into the empty spaces in the abomasum and when this happens the troubles really start (*diagram 7*).

The fibre irritates the inside lining of the abomasum and leads to indigestion. The same irritation can produce a calf fit, which is often fatal.

Where the irritation is excessive, an ulcer may form; this, together with the fermenting milk in the rumen, leads to bloat — another nuisance that can be fatal. Of course the bloat can and mostly does occur without the ulcer.

But, worst of all, the fibre may block up the very narrow exit from the abomasum into the duodenum. When this happens, the calf receives only a fraction of its available nourishment and literally starts to fade away suffering from every kind of protein, carbohydrate and vitamin deficiency until eventually it goes off its legs and dies.

At the same time, the lowered resistance caused by all these troubles leaves the calf an easy prey to scours and all the other killer diseases.

This, in my opinion, is the simple explanation of why so much money is lost in calf rearing. It is because the undigested fibre gets into the empty spaces in the abomasum, empty spaces that shouldn't be there. What action should be taken? Here's what I advise.

ADVICE TO FOLLOW

First of all, there is the matter of colostrum. A calf ideally requires 2½ litres of colostrum in divided feeds during the first four to six hours of life. The colostrum protects the calf from all the local microbes and without the colostral antibodies the calf has no natural resistance against disease.

The calf is capable of taking up its full quota of colostral antibodies **only when it is around four to six hours old**, so if you are lucky enough to see the calf being born, milk off 2-2½ litres of colostrum four hours later (*photo 1*) and drench the calf with it or give it by stomach tube.

A better idea is to let the calf suck the mother for at least 24 hours, and the best idea of all is to rear the calf on the cow.

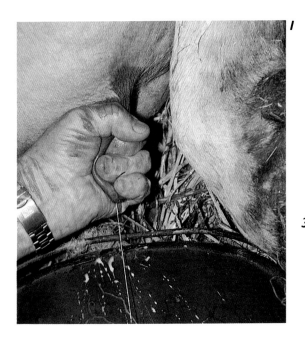

1 If you haven't got an artificial suckler, then for the first 14 days of the calf's life suckle it three times a day. This will mean three clots daily in the abomasum and will leave less room for the irritating fibre.

Use the mother's milk for the first week and bulk milk for the second, introducing milk substitute only towards the end of that period. Give the milk in a bucket with a teat at the base and hold the teat downwards to simulate the cow (*photo 3*).

3

If you can't suckle your calves on cows, then use an artificial suckler, which ensures that the calf's head is held at the correct angle and that the oesophageal groove functions properly. Also, with the artificial suckler the calf keeps helping itself as it would do on a cow and the abomasum is kept full of small clots (*photo 2*).

If the calf refuses to drink, then get your veterinary surgeon to give the colostrum by stomach tube.

Water is absolutely vital. Without water rumenal digestion cannot begin to take place because digestion in the rumen is a fermentative process. But the water must be clean and fresh (*photo 4*).

2

4

For fibrous food provide best quality hay fed at head level from the second or third day (*photo 5*), and corn from the third or fourth day.

If you do all this, then you can wean your calves suddenly, without any losses or troubles, towards the end of the sixth week. At this age the calf's digestive system is fully developed and ready to cope with normal bovine food.

But remember: It's not much use developing perfect feeding techniques and then keeping the calves in draughty death-trap houses with soaking wet bedding. Don't forget to give them draught-proof kennels and warm dry beds (see 'Calf Housing', page 80).

To sum up: Get as close to nature as you can and you'll minimise your calf losses.

FEEDING THE BOUGHT-IN CALF

If the purchased calf looks in any way under the weather, it should be put in the hospital pen for a few days. If it is apparently healthy and vigorous, then it can go straight into the rearing pen. But in any case give colostrum by stomach tube if in any doubt at any age up to six weeks.

For the first *two* liquid feeds, the calf should be given at each feed 2.5 litres of water (at blood heat) containing 60g of

5

powdered glucose. Ad lib hay should be available, but no nuts or additional water. After 24 hours, put the calf straight onto milk or milk substitute, depending on its age or system of feeding, and introduce the nuts and ad lib water.

If this extremely simple routine is followed exactly, provided the calf pen floor and environment are up to scratch, losses and diseases should be kept to an absolute minimum.

21
Bloat in Calves

THE SUCKLED CALF

As already explained in 'Calf Feeding', page 87, the bucket-fed calf is the most likely to suffer from bloat, chiefly from fermenting milk which

gets into the rumen when there are no digestive enzymes there to deal with it.

Symptoms
In severe cases the calf shows typical signs of

colic — kicking at the belly and flopping down. Scour may or may not be present.

However, in my experience most cases are mild. Nonetheless the patient is dull and unthrifty and it is wise to consult a veterinary surgeon (*photo 1*).

rumen with a large bore hypodermic needle (*photo 3*). He may then inject a muscle relaxant intravenously (*photo 4*), and prescribe an oral antibiotic. He will advise removing all solid food for several days to allow the rumen to empty completely.

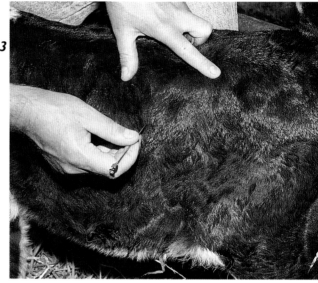

Treatment

The veterinary surgeon will relieve the bloat either by passing a stomach tube down the calf's throat (*photo 2*), or by puncturing the

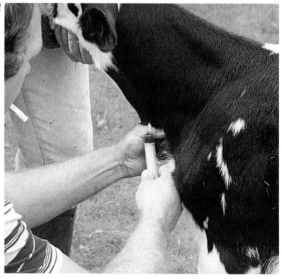

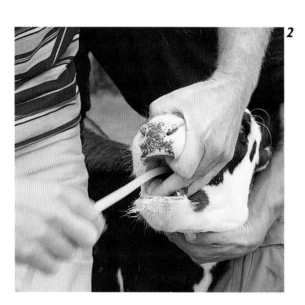

If the patient is scouring he will probably prescribe two or three days on an electrolyte solution before returning to full milk or milk substitute.

The Weaned Calf

This is due mainly to the rumenal contractions not functioning properly and is difficult to cure.

Treatment
Again, this is a job for a veterinary surgeon who will deal with the case similarly to that in the suckling calf though he will most likely pass the stomach tube up the larger calf's nostril (*photo 5*) and advise a 10 day return to a liquid diet plus the removal of concentrates during that period.

22
Calf Scour

There are four common types of calf scour: digestive (or non-specific) scour; E. coli (or white) scour; Rotavirus and/or Coronavirus scour; and the dreaded salmonella infection.

Research has shown the Rotavirus is by far the most common cause of calf diarrhoea, with Coronavirus second in importance. Contrary to earlier opinion, E. coli is only a danger during the first few days of the calf's life and not throughout the entire suckling period.

Digestive Scour

Cause
Digestive scour occurs mainly when a calf is sucking on a cow which is giving too much milk (*photo 1*). The diarrhoea is due simply to an excess of milk, which passes straight through the stomachs into the calf's small intestine.

Similar scour can be produced by bucket feeding to excess, but this is much less common.

A simple digestive scour occasionally arises as a result of too rapid changes in the diet, irregular feeding, or feeding too hot or too cold, but in such cases the picture is usually complicated by a flare-up of E. coli or more likely the Rotavirus infection.

Symptoms

Whenever you have scour occurring in a calf or calves that are sucking on a cow, then the condition is a simple digestive scour in 99 cases out of 100. The dung is usually yellow or yellowish-white in colour (*photo 2*).

3

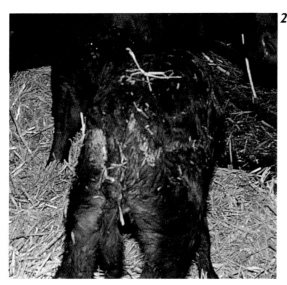

2

Prevention

Obviously, it is unwise to put a single calf on a cow giving a lot of milk. The general rule I recommend is one calf per three-quarters of a gallon of the cow's milk.

By their consistent sucking, the individual calves will stretch this estimated three-quarters to at least the full gallon. I know many successful calf rearers who allocate one sucking calf per half gallon — certainly it is better that the calf should have to work for its ration (*photo 4*).

In other words, the best way to prevent digestive scour in calves is to get back as near as possible to nature.

4

Treatment

Take the calf off the cow for 24 hours. During that time give two feeds each of 2½ litres of warm water containing 60-65g of powdered glucose and a whisked raw egg (*photo 3*). Thereafter, for the next three days, restrict the calf's sucking to three minutes three times a day.

This simple treatment will cure digestive scour most effectively. If, however, the scouring persists, it often indicates that bacterial complications have set in and it may be necessary to proceed with the routine prescribed for E. coli scour.

E. COLI INFECTION OR WHITE SCOUR

This is seen only in the first few days of the calf's life (*photo 5*).

Cause

One specific cause is the dangerous strain of E. coli bacterium (known to scientists as K99). E. coli is a normal resident of the intestines of practically every calf. The calf picks the germ up by the mouth after it is born.

Predisposing factors

Any factor which lowers the resistance of a calf to infection can predispose to white scour by allowing the E. coli to multiply and grow strong. The common factors are:

- Lack of colostrum, so that the calf has no antibodies and consequently no resistance against E. coli.
- Exposure to rapid changes in environmental temperature (see 'Calf Housing', page 80).
- Cold, wet floors and bedding.
- Transport to markets (*photo 6*) and exposure for long periods in the market pens (*photo 7*).
- Irregular feeding, overfeeding and feeding at an inconsistent temperature.
- Keeping calves continually in the same box.

Here the germs increase in number and become progressively more dangerous.

Symptoms

White diarrhoea (*photo 8*) associated with a fall in body temperature; the calf's ears and tail become ice-cold to the touch.

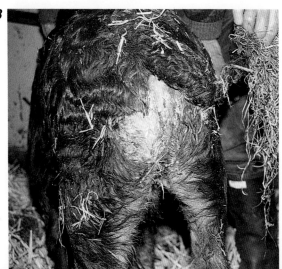

95

Treatment

Constant warmth is the first essential, so the patient should be immediately put under an infra-red lamp in a hospital pen, improvised or otherwise — see 'The Hospital Box', page 85. This is the most vital and important part of all scour treatment.

Milk or milk substitute should be withheld for 24 hours and warm water, raw egg and glucose used, as in digestive scour. The chief cause of death is often dehydration so intravenous fluid therapy is vital.

Swabs of faeces should be taken by your veterinary surgeon and sent to a laboratory immediately. There the E. coli can be identified and drug sensitivity tests can be done inside 24 hours. This allows the veterinary surgeon to apply the most effective specific treatment.

Pending this laboratory investigation, a general broad-spectrum antibiotic combined with a bowel sulpha drug and an anti-dehydrating agent should be given by the mouth (*photo 9*).

Vitamin injections, iron tonics and intravenous saline injections can also be used but these should be prescribed and given only by the veterinary surgeon.

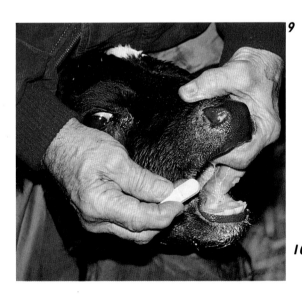

9

Prevention

The obvious answer is to eliminate, as far as possible, the predisposing causes. Therefore, the routine should be as follows.

- Make certain the calves get their own mother's colostrum or first milk for at least the first five days of life, but especially in the first four to six hours after birth.
- Get the floors, housing and environment to the standards specified in 'Calf Housing', page 80. Especially important is a constant temperature.
- Avoid buying calves from markets if possible.
- Pay special attention to regular and correct transitional feeding as recommended in 'Calf Feeding', page 87.
- Every three to six months, empty the calf pens completely. Scrub with hot water, soda and disinfectant and leave completely empty for at least 14 days. This will eliminate the risk of disease build-up.
- Vaccines against E. coli are variable in effect. Be guided by your veterinary surgeon.

Do all this and you will get little, if any, scour or losses in calves. Don't do all this and you deserve all the trouble and deaths you will most certainly encounter.

ROTAVIRUS AND/OR CORONAVIRUS SCOUR

This, the most common type of calf scour, flares up usually during the first 10 days of life. The viruses gain entrance to the patient in faecal-contaminated food.

Rotavirus and/or coronavirus scour is a contagious condition which can spread rapidly through an entire calf unit.

Symptoms

A sudden onset of a watery yellow diarrhoea (*photo 10*) followed quickly by depression,

10

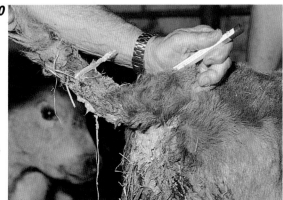

dehydration and eventually prostration with the eyes sunken and the inside of the mouth cold and clammy. It is the dehydration that causes death if veterinary treatment is not promptly applied.

Treatment
Immediately isolate to prevent the spread of infection (*photo 11*), and then call the veterinary surgeon as quickly as possible.

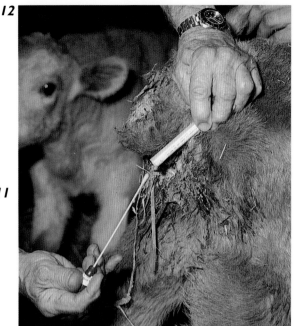

Such cases, after only one day scouring, require nearly 4 litres of fluid replacement. The veterinary surgeon will give a balanced electrolyte salt solution intravenously and by stomach tube. He may have to siphon off toxic fluid from the calf's stomach and replace it with the warm electrolyte solution. Certainly if the correct treatment is given promptly the result can be spectacular. Any delay, and death is a near certainty. Antibiotics and scour tablets or drenches are of no use whatsoever in virus infections. However, it is wise to get the veterinary surgeon to send a faeces sample to a laboratory for examination in case secondary bacteria are involved — salmonella particularly (*photo 12*).

Prevention
Obviously it is much better to try to prevent this severe type of calf scour by tightening up management and my advice is to follow the procedure outlined for the control of E. coli with a heavy emphasis on establishing passive immunity, which is obtained entirely via the colostrum.

Fortunately an efficient vaccine is available and it is economically sensible to use this as a routine.

SALMONELLA SCOUR

Cause
Chiefly a germ called *Salmonella dublin*, although another member of the same group — *Salmonella typhimurium* — can also cause a great deal of trouble.

There are five sources.

- Infected calves.
- Recovered cows.
- Pigs, poultry and human-beings which are infected by *Salmonella typhimurium*.
- Rats.
- Infected buildings, transport vehicles and markets. **Salmonella can live in a dirty building for several years.**

How it spreads

Just one single infected calf from a market, a transport vehicle or from a dealer's premises, is all that is necessary to create havoc. In other words, salmonellosis is a disease which is spread chiefly by bought-in calves (*photo 13*).

It is usually fatal in adult cattle. But *Salmonella typhimurium* infection can frequently flare up in pigs and poultry and the germ can be carried to the calves via the feet and clothing of the attendants.

Human infection is also on the increase.

Salmonella, once introduced, can affect the healthiest of calves but, nonetheless, the poorer the management the more likely the infection is to persist and gain strength. Thoughtless and improper feeding, cold and draughty pens, wet or inadequate bedding — one of these can lower the calf's resistance to such an extent that it stands little chance of survival.

If you add to that periods of starvation in cold, draughty markets and jolting transport journeys, it is easy to understand why salmonellosis can become an endemic problem.

How it differs from other causes of scour

It appears to attack chiefly the older calves — three, four or up to six weeks old — an age beyond the usual calf scour period though, of course, young calves are also highly susceptible. The average calf receives, in the colostrum, antibodies against E. coli scour but only rarely against salmonella. Consequently, when salmonella is introduced the results can be disastrous, with a death rate of up to 50 per cent.

13 ## Symptoms

The germs cause an inflammation of the stomach and intestines, and give rise in calves to an elevated temperature of around 40°C (105°F), with a diarrhoea varying from yellow to dark-coloured and bloodstained, depending on the degree of severity (*photo 14*).

The illness runs a course of five to six days, but in hyperacute cases the calves may die in from 24 to 72 hours, sometimes with little or no signs of scour. Occasionally the picture is complicated by the development of a well-marked pneumonia.

14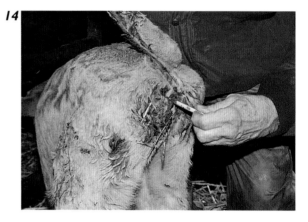

Treatment

A veterinary surgeon should be called in immediately. He will take swabs of the faeces (*photo 15*) and will culture and identify the germ in a laboratory.

15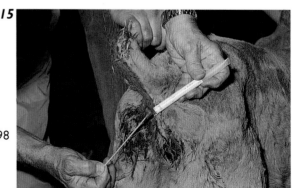

Mortality is about 50 per cent, despite the wide range of drugs at our disposal. Any drug prescribed should be used in conjunction with saline and vitamin injections and simple old-fashioned remedies like kaolin, chalk, stout, eggs and brandy.

Obviously, therefore, with salmonella, it is much better to concentrate on prevention.

Prevention

A salmonella vaccine is available and this should go a long way to reducing the incidence. However, to be completely effective, the vaccine use must be combined with commonsense husbandry, if only because the vaccine takes 14 days to produce its protection.

If you've had the infection, clean the box out very thoroughly with hot water and soda, disinfect, and leave empty for at least 14 days before restocking with vaccinated calves. At the same time, organise an all-out attack on the rat population.

If you haven't had the infection, aim at keeping the herd self-contained. If unable to do this and you have to buy in calves, watch your management carefully.

First, if possible, purchase calves from known disease-free sources. I know farmers who travel hundreds of miles to collect calves simply to make sure of their origin.

Second, with every newly purchased calf get your veterinary surgeon to examine it (*photo 16*) and keep the calf in isolation for at least 14 days.

Third, make sure your calf pens are clean, dry, warm and draught-proof (see 'Calf Housing', page 80).

And, last, watch the feeding very carefully during the settling-in period (see 'Calf Feeding', page 87). Any stomach or bowel upset can lower the calf's resistance and allow salmonella to gain hold. It is often during the transition

16

period of dietetical change that disease flares up.

I cannot emphasise too strongly that the better the calf's condition, feeding and environment, the less likely it is to succomb to any disease.

N.B. In recent years salmonellosis has become an ever-increasing problem in *adult* cattle (see my other book, *Cattle Ailments*). Vaccination of adults does give some immunity to the calves.

BVD IN CALVES

The virus of bovine viral diarrhoea can, very occasionally, produce severe scouring in a calf since it has been shown that an infected mother can transmit the virus to her offspring during pregnancy. Obviously such a condition is extremely difficult to diagnose and treat and it is recorded here merely because it can occur.

23
Coccidiosis

Cause
Coccidiosis is caused by a minute parasite belonging to the *Eimeria* family.

Source
Calves are infected by adult cattle, many of which are carriers of the parasite without showing any symptoms. The parasite is present in the carrier cow's dung.

Symptoms
Eimeria produces its effect on susceptible calves by burrowing into the wall of the large intestine. This irritation makes the calf strain intermittently and persistently to pass dark-coloured loose faeces. The dark colour is due to the presence of blood in the faeces. There may also be shreds of the intestinal lining present (*photo 1*).

If untreated the calf rapidly loses weight and may die.

Treatment
Get a veterinary surgeon to confirm the diagnosis either from experience or by examining the faeces microscopically. He will prescribe the appropriate treatment, the most popular being a course of an oral sulpha drug plus a cover injection of long-acting broad-spectrum antibiotic to control possible secondary bacterial infections (*photo 2*). No doubt future research will produce other specific drugs.

Prevention
Since the infected calf can rapidly contaminate others it should be isolated immediately and all incontacts dosed with the oral sulpha drug, then removed to clean, fresh quarters.

The coccidiosis parasite can exist for a long time and is highly resistant to ordinary disinfectants. Consequently, the infected pen or box will require special cleansing under the supervision of a veterinary surgeon.

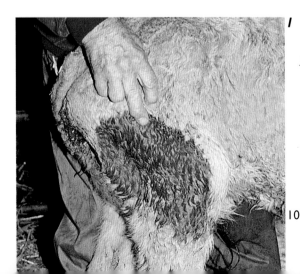
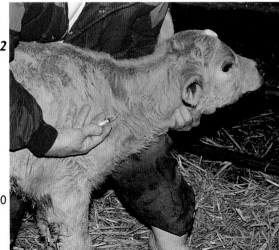

24
Calf Diphtheria

Calf diphtheria is a simple condition which, in my experience, is little understood by the average farmer or stockman. It is a common disease occurring everywhere and is one of the easiest of all calf diseases to diagnose and treat.

Cause

Calf diphtheria is caused by the same germ that causes foul in the foot — *Fusiformis necrophorus*.

It is a normal resident of the feet of a high percentage of cattle, living in the cracks between the wall and the sole (*photo 1*). It can live in the dung and the bedding, but only for about a month at the most. Whilst outside, it can live on the pasture for a maximum of 14 days.

The germ gets into the calf's mouth simply through the calf eating contaminated bedding or contaminated fodder. It gets a hold through wounds, scratches or cuts in the linings of the mouth cavity. The scratches can be caused by straw, barley piles, thistles or thorns.

As soon as the germ gains entrance into the wound, it multiplies and grows and produces a lump of black dead tissue, in exactly the same way as it does in a case of foul (*photo 2*).

Symptoms

During the multiplication of the germ, the affected part of the jaw becomes painful and the calf shows a disinclination to eat or take its suckling. The temperature is usually normal but the patient is dull and listless and is not keen to stand or move around.

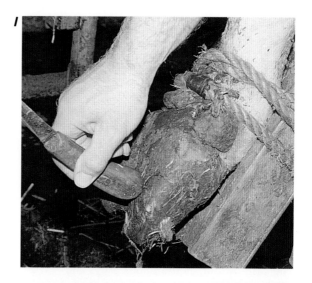

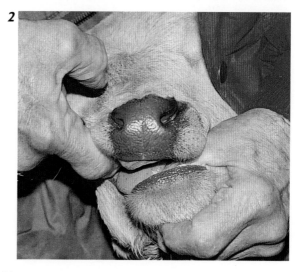

101

Later on, the only symptom is a lump on the side of the jaw and, in the majority of cases, this is the first symptom seen (*photo 3*).

Occasionally, the germ attacks the soft tissues of the tongue and the back of the throat. When this happens, the outstanding symptom is an inability to swallow. Boluses of partially chewed hay are found in the back of the mouth and the calf's breath smells like dead meat. Naturally, condition is rapidly lost.

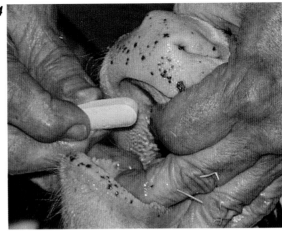

4

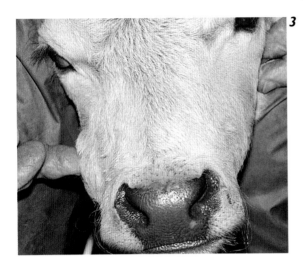

3

Treatment

Fortunately, treatment presents no problem, provided it is started in reasonable time. As in the case of foul, all the sulpha drugs and antibiotics are virtually specific to the bacterium. When the sulpha drugs or antibiotics are given by injection, then one dose of a long-acting antibiotic is sufficient. When given by the mouth, a four- or five-day course is required (*photo 4*).

Prevention

There is no specific vaccine against *Fusiformis* (apart from that used against foot rot in sheep) and, therefore, once again the best means of defence lies in cleanliness and good husbandry.

In the first place, calves should be housed in a box which has been cleaned, disinfected and rested for at least 14 days before their entry.

Water ad lib should be provided from birth and ample hay should always be fed from a rack or net. It is my experience that calves will rarely if ever eat straw when good hay is available.

Wet bedding seems to favour the persistence of the germ and, therefore, the floor underneath the straw should be insulated and well drained or, if not, it should be covered by a good thick layer of ashes.

Where the disease is a special problem, it may be a good plan to scrape the cracks in the calves' feet thoroughly with a blacksmith's knife and afterwards to soak them individually in a 10 per cent solution of formalin, or some other mild antiseptic.

Many people say that calf diphtheria is only found where the husbandry is bad. I wouldn't go to the length of saying this, but I would suggest that it is much less likely to occur where the husbandry is good. It is not a fatal condition, but it is a disease which every good stockman should know about.

25
Joint-ill, Infected Navel, Navel Abscess and Umbilical Hernia

Joint-ill usually affects young calves round about a week or ten days old although occasionally the symptoms may appear at three weeks or even a month (*photo 1*).

germs which cause scour and pneumonia. In other words, a disease build-up occurs (*photo 2*).

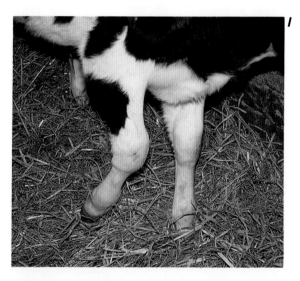

Cause
Joint-ill in calves may be caused by a variety of germs, the commonest being simple *Streptococcus* which can be present almost anywhere.

In a pen where calves are continually housed, germs appear to get progressively more dangerous and the number of cases of joint-ill will increase. This applies also to other

Entrance and effect
The joint-ill germs can gain entry through an abrasion or wound but usually they get in via the calf's navel during the first two or three days of life (*photo 3*). Just occasionally, joint-ill can flare up as a complication of some other debilitating disease.

The germs get into the bloodstream and are carried to the various joints where they

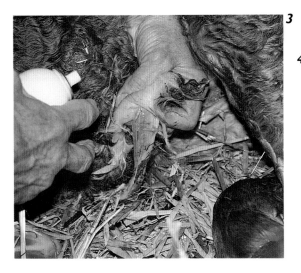

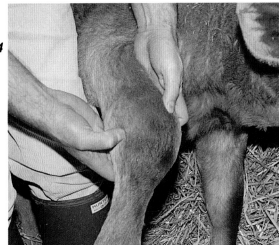

multiply and produce inflammation and swelling, and eventually pus if allowed to progress without treatment.

Symptoms

The affected calf or calves — often there is more than one affected — start to walk stiffly. At the same time they go off their feed and run a high temperature of 40.5 – 41°C (105° – 106°F). The navel is usually swollen and there may be a painful puffiness in one or several of the joints (*photo 4*).

In long-standing cases, abscesses form in the joints (*photo 5*) and these may burst externally.

Treatment

Fortunately, there are sulpha drugs and antibiotics available which are specific against the common types of joint-ill. But the drugs should be used carefully and only under veterinary supervision (*photo 6*).

Prevention

Scrub, disinfect and rest the calving boxes for 14 days at least twice a year; and scrub, disinfect and rest the calf pens at least four times a year. This breaks the cycle of disease build-up.

Use plenty of clean bedding in the calving box during and after each calving, and subsequently in the calf pens when the calves are transferred there.

Dress the navel of the new-born calf with an

antiseptic or antibiotic dressing three times on the first and second day and twice on the third day of life (*photo 7*). This precaution is especially important when the calves are taken from the mother at birth. An aerosol containing a powerful antibiotic and dye is ideal for this purpose, though sulpha powder dressings can be used, and modern iodine dressings are also excellent.

Lastly, and perhaps the most important point of all, the new or recently born calf should be put in a pen with a clean dry bed (see 'Calf Housing', page 80).

Treatment
Long-acting broad-spectrum antibiotics as prescribed by the veterinary surgeon are used to treat infected navels.

Prevention
Prevention is identical to that recommended for joint-ill, paying particular attention to the calf's navel as soon after birth as possible.

Antibiotic aerosol sprays are effective as are iodine preparations (*photo 9*).

9

7

INFECTED NAVEL

Infected navels not only predispose to joint-ill but the infection may extend into the abdomen causing fatal peritonitis (*photo 8*).

NAVEL ABSCESS

When an abscess develops (*photo 10*) surgical lancing under a general anaesthetic by a veterinary surgeon is necessary. The reason? The abscess may conceal a hernia and lancing by a lay person could prove disastrous.

8

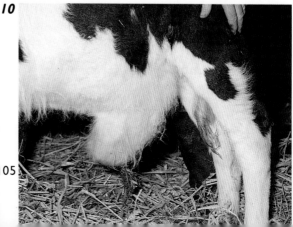

10

UMBILICAL HERNIA

Umbilical hernias are reasonably common in calves (*photo 11*). The majority can be successfully operated on by the veterinary surgeon (*photo 12*).

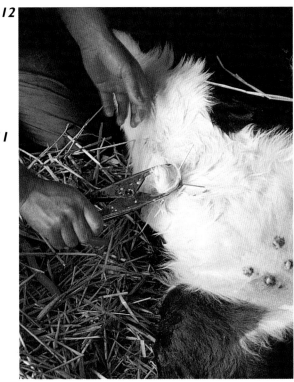

26
Contracted Tendons

Occasionally a calf is born with one or both forelegs bent forward (*photo 1*) and when it stands, the fetlocks and/or the knees knuckle forward. Such calves are often scrapped because of the apparent deformity.

It is wrong to slaughter such calves because the vast majority will get better by themselves within the first two months of life.

The condition is due to a contraction of the flexor tendons which run down from the back

of the knee joints to the bulbs of the heels (*photo 2*).

Even though, in the beginning, it seems difficult or impossible to straighten the affected leg, nature appears to relax the tendons gradually. One important point, however, is that when the calf is born with the legs straight and the contraction develops subsequently, the condition is much more serious and rarely, if ever, gets better.

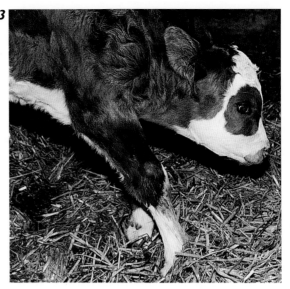

3

2

FRACTURES

Broken legs in calves usually heal quickly, particularly with the aid of fibre glass bandages (*photo 4*).

4

The only treatment required is to keep the calf on a deep, soft bed of built-up muck topped with straw (*photo 3*). If this is not done, the front of the fetlock joints may become lacerated or even worn through to the bone.

Just a few days or weeks of patience and care and the result is a healthy straight-legged calf growing into money.

27
White Muscle Disease

The cause of this disease is a deficiency of vitamin E or selenium. Most purchased farm rations contain ample amounts of both these though many grazing pastures may be deficient. Nonetheless symptoms are most likely to appear if the dams have been poorly fed and the calves have been reared on skimmed milk without access to good hay. Another possible cause is the feeding of excess cod-liver oil.

Symptoms
The first symptom of white muscle disease in the calf is sleepiness combined with an unwillingness to stand. When made to stand, the calf adopts unnatural postures and day by day the muscles around the top of the hind and forelegs and in the rest of the skeleton waste away rapidly (*photo 1*). If the calf is untreated, it may finally become completely prostrate, show distressed abdominal breathing, and die within a few days. If the calf is excited, it may fall down and die from a heart-attack. In fact occasionally this is the first sign of trouble due of course to the disease affecting the heart muscle.

Treatment
It is important to start treatment at the earliest possible stage. If the condition becomes advanced, there is no cure.

As soon as muscular weakness and degeneration is suspected, vitamin E in the form of tablets or wheat germ oil should be fed in small quantities daily (*photo 2*). Also, concentrated vitamin E and selenium injections

should be given by your veterinary surgeon.

Combined with this medicinal treatment, ample good quality hay should be provided for the calf and, if possible, milk should be used for suckling, although most good quality milk substitutes contain adequate vitamin E and selenium. If cod-liver oil is being fed, it should be stopped immediately.

Prevention

Prevention is simple. It is merely a matter of good management and correct feeding.

The dam should be well fed, particularly during the latter half of pregnancy and should, in particular, have an ample supply of good hay or silage. Where the disease appears in silage-fed animals, then hay should be added to the dam's diet during the last two months of pregnancy.

Cod-liver oil should not be fed to calves, except under veterinary supervision. Small amounts of wheat germ oil (approximately 2-5g) can be fed to the calf daily for the first 14 days of life. This is an effective but expensive preventative and is not usually necessary.

28
Vitamin A Deficiency

Vitamin A is the essential growth vitamin. Without it, there is first of all a retardation and then a cessation of growth in all the body bones, muscles and tissues.

Normal sources

Vitamin A is stored in the calf's liver. The chief natural sources for the young calf are first of all the mother's milk and later the bulk feed of hay, silage or haylage. The other usual source is provided by the vitamin-supplemented proprietary calf-rearing foodstuffs (*photo 1*).

Causes of deficiency

Vitamin A deficiency is most likely to arise in weaned calves, under normal circumstances, if they are being fed poor quality fibre and unsupplemented home-grown cereals (*photo 2*).

Excess feeding of barley when little or no hay is provided can precipitate severe attacks. Here the condition is produced, at least in

109

part, by the widespread liver damage that so often occurs with intensive barley feeding.

Symptoms
General unthriftiness and failure or cessation of growth. The simple reason for this is because, as mentioned in the introduction, vitamin A is the growth vitamin, and when it is deficient or absent, all growth is correspondingly affected. The hair falls out (*photo 3*) and later the calf may stagger slightly when it moves.

3

There occurs a thickening of the skin (*photo 4*) and, later, partial or complete blindness, with the eyes occasionally discharging.

4

Treatment
If blindness is a symptom, it is usually permanent and it is best to slaughter the calf and concentrate on preventing symptoms appearing in the others. If the patient is a bull calf being reared for breeding, then it should be slaughtered or castrated since its fertility is likely to be permanently damaged.

In addition to providing a natural or supplemented vitamin A source, 250 to 500 thousand international units of concentrated vitamin A should be injected intramuscularly into all the remaining calves. Recovery will not be spectacular — it will be slow but progressive.

Prevention
The simple prevention, apart from sticking as closely as possible to traditional rearing methods, is to provide ad lib top-quality early-cut hay from early calfhood (*photo 5*) or high-quality silage or haylage. In intensive barley beef production the good quality fibre may be rationed, but a percentage is nonetheless essential.

5

29
Lead Poisoning

All cattle, especially calves, are highly susceptible to poisoning by lead (*photo 1*). A comparatively small quantity is enough to kill a calf and many of the mysterious outbreaks of illness and death among groups of stock in various parts of the country have eventually been traced to lead poisoning.

personally have had to investigate have been due to the animals licking or eating one or the other of these.

Roofing felt, in itself, does not contain lead but it is often impregnated with lead paint. Felt may flake from the layer underneath the tiles and fall onto the food, or it may have been used to patch up holes in walls or windows and attract the calves during their idle exploratory licking.

In most of the felt cases I have seen, portions — usually from old fowl pens — have been left lying about in the yard and have found their way into the calf boxes either in the bedding or in the hay (*photo 2*).

Sources of lead
The most common sources are **paint** and **roofing felt**. In fact, all the cases which I

Many modern paints do not contain lead, but all the older paints did and the basic paint work inside many older farmsteads is the source of great potential danger.

Only a few flakes of this old paint are required to produce symptoms in calves (*photo 3*) and yet, in many cases, calves have been reared without loss year after year in the same boxes with identical paint on the doors. This is simply because it takes a long time for a good-quality paint to start peeling, and a calf will rarely persist in licking a hard, smooth surface.

One of the chief sources of lead paint — a source that is often forgotten — is the iron girders that so often form an integral part of the low roof of a calf pen. Usually such girders are coated with a preservative layer of red lead paint and, when this starts to flake, it falls into the troughs or on to the hay and straw.

In the less acute case, the calf stops eating, shivers and becomes ice-cold, especially at its extremities. Often it stands with its head pressed against a wall (*photo 5*). It may stumble and stagger and flop just like a hypomagnesaemia case, though close observation will show that the unsteady movement is mostly due to a loss of vision. In other words, blindness is a diagnostic feature of the condition.

5

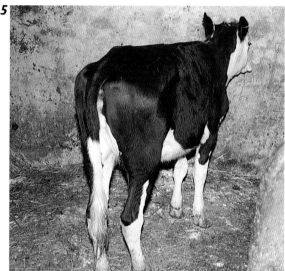

In recovered cases, blindness may persist for several months. In fact, I have seen it persist for two years in feeding cattle and, occasionally, it can be permanent (*photo 6*).

3

Symptoms
In some ways, the symptoms of lead poisoning are similar to those of hypomagnesaemia. In fact, in acute cases they are almost identical (*photo 4*).

4

6

112

In fatal cases, death is usually preceded by an epileptic-like convulsion (*photo 7*).

9

7

There is a more specific antidote, but this has to be given intravenously and should only be administered by a veterinary surgeon.

It is wrong to panic into the emergency slaughter of a blind animal, particularly one that can be fattened for the butcher, because it is almost uncanny how a blind animal adapts itself to a normal life, finding both trough and water bowl by instinct.

Treatment

Any form of treatment is highly unsatisfactory. The toxic effect of the lead often leaves the animal's liver and/or kidneys irrevocably damaged and the blindness, as mentioned above, may be permanent or at least take a long time to get better.

The best first-aid antidote is Epsom salts (i.e. magnesium sulphate). In lead poisoning, Epsom salts produces its effect by combining with and inactivating the lead.

The required dose of Epsom salts for a calf is two teaspoonsfuls dissolved in water and given as a drench three times daily for at least five days. For an adult animal, up to 113g (4oz) can be given twice daily (*photos 8 & 9*).

Prevention

So often it is a matter of shutting the stable door after the horse has bolted out. The answer to lead poisoning lies in simple commonsense and good management.

Since treatment is an unsatisfactory compromise, the real answer lies in making sure that you don't get lead poisoning in the first place. This is easy. All you have to do is to check your buildings, calf pens, yards and pastures for possible sources of lead.

In calf pens, particularly in older buildings, a priority job must be to burn the paint off any supporting or roof girder and re-coat with non-lead aluminium paint. At the same time, blow-lamp, scrape and repaint all iron beams and inside doors, again with a lead-free compound (*photo 10*).

8

10

I would say that this should be done in all pens when it is not definitely known whether or not the paint contains lead. The job is not difficult or expensive; it is merely sensible and well worthwhile. The important thing is to make sure from the supplier that the new paint does not contain lead.

Make absolutely certain that none of the herd has access to rubbish dumps or felted fowl pens.

30
Dehorning

The dehorning of calves is a task which can be done by any efficient farmer or stockman, but it is essential that the correct technique is employed. According to UK law an anaesthetic must be used. General anaesthesia is dangerous and out of the reach of the farmer, but local anaesthesia is simple and most veterinary surgeons will be only too happy to instruct their clients in its use.

Calves should be dehorned when under one month old. Usually around two weeks old is best as by then the horn bud, though easily distinguishable, is not too well-developed (*photo 1*).

The dehorner
All caustic preparations are dangerous and unreliable. The job should always be done with a correctly heated dehorning iron.

Electrically-heated dehorning irons are excellent but I have found that the most satisfactory all-purpose instrument is the one heated by calor gas, because calves are often kept in boxes where there is no electricity supply. The calor gas dehorner is efficient, easily transported and comparatively inexpensive (photo 2).

A portable battery-powered dehorner (*above photo*) fuelled by lighter fuel is an alternative.

1

2

Other tools required

Local anaesthetic, syringe and needle, scissors, antiseptic and cotton wool (*photo 3*).

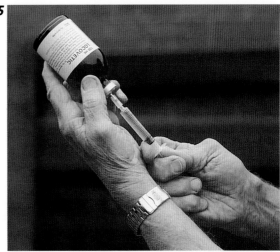

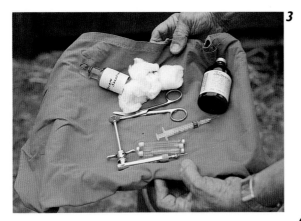

A needle 18mm long, i.e. the one on the left (*photo 4*), is ideal, although one with a shorter and thicker bore can be used provided the point is sharp. The needle should be at least 12mm long because, in calves, the nerve to be blocked lies fairly deep at the site of injection.

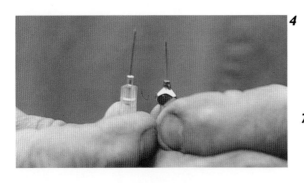

An ordinary record syringe is ideal and local anaesthetic can be supplied in bulk (*photo 5*).

A special cartridge-loading syringe can be bought which, though expensive, is perhaps more economical in the long run (*photo 6*). The local anaesthetic is supplied in cartridges and one cartridge contains sufficient to anaesthetise two horn buds.

The job can be done comfortably with the calf on its legs and its hindquarters in a corner (*photo 7*).

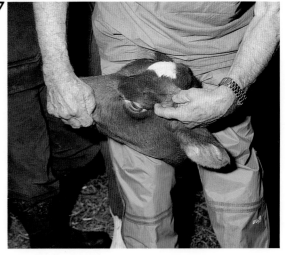

One or two assistants can keep the hind end in position, though one person is enough; in fact, the job can be done single-handed if necessary. The operator should stand astride the shoulders of the calf, holding it firmly between his knees.

Technique

Dealing with the right horn bud, the operator should hold the calf's head firmly across his thigh with his left hand. With the scissors he should now clip the hair over the horn and over the site of the injection (*photo 8*).

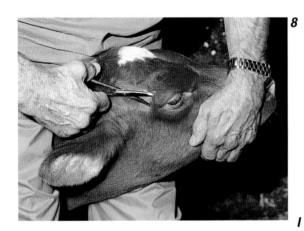

The site for injection of the anaesthetic is a spot about mid-way between the eye and the horn bud, and immediately below the ridge of bone which runs between these two.

The site should feel soft and pliable under the finger (*photo 9*).

After the hair is clipped, the site should be coated with a powerful skin antiseptic (*photo 10*).

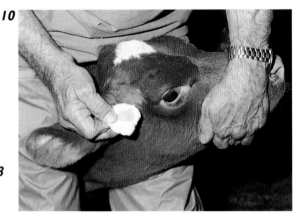

The injection

The needle should be inserted fairly deeply and at right angles to the ridge. One cc (i.e. half a cartridge) of local anaesthetic should be injected (*photo 11*). Inexperienced operators should inject 2-4 cc.

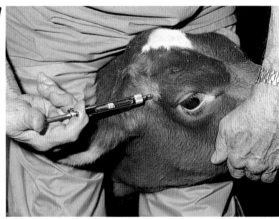

One sure way of telling whether the needle is in the correct spot is by the ease with which the anaesthetic goes in. If the syringe plunger has to be forced, the needle should be withdrawn and inserted again slightly lower down.

The clipping, disinfecting and injecting should now be repeated on the left side, this time the calf's head being held firmly against the right thigh of the operator by the right hand of the assistant.

It takes approximately five minutes for the anaesthetic to take full effect, an effect which lasts for at least one hour, so it's a good idea to anaesthetise all the calves that require dehorning before commencing the actual operation.

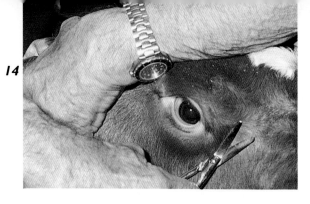

14

The dehorning
The end of the dehorning iron should be placed over the bud and turned in a half circle 10 or 12 times (*photo 12*).

15

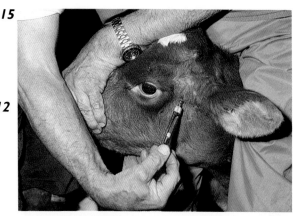

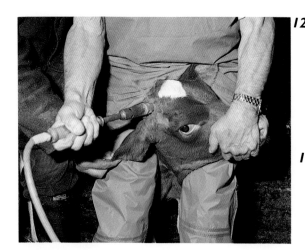

12

The horn bud should now be dug out by pressing the upper edge of the iron firmly inwards and downwards.

Evidence that the job has been done correctly, and that the horn will never grow again, is the horn bud in the end of the dehorning iron and a clear hollow in the skin of the head where the bud used to be (*photo 13*).

The left horn bud is now removed in exactly the same way (*photos 14 – 17*).

16

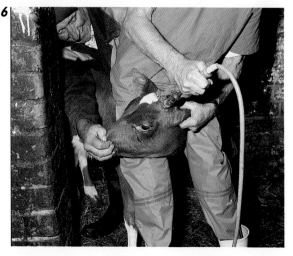

17

13

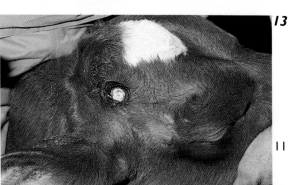

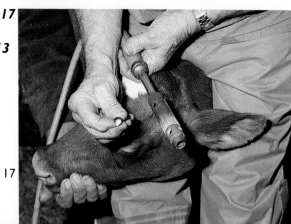

31
Removal of Accessory Teats

Nothing is more unsightly than an adult dairy cow with more than its quota of teats (*photo 1*).

easy handling, the younger the calf the better; I think it should be under two months old (*photo 2*).

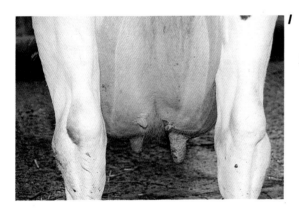

I've seen many honest attempts by herdsmen to remove accessory teats in calves but rarely, if ever, have I come across a perfect job.

The most common mistake is for the teat to be cut off level with, or even slightly below, the surrounding skin. This leaves a lump or scar which gets more and more unsightly as the heifer grows. And if the amputation is attempted when the heifer is older, a milk fistula may result, i.e. a permanent hole which will leak milk after the heifer has calved.

There is, however, a simple technique which always ensures a first-class job.

First of all, get an assistant to turn the calf up and sit it on its hind end. Obviously, for

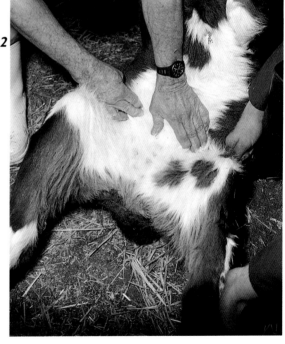

The tools required for the job (*photo 3*) are a looped cord, a pair of sharp curved surgical scissors, a pair of artery forceps, a small syringe, some local anaesthetic, and a tube of antibiotic. All these can be provided by your veterinary surgeon.

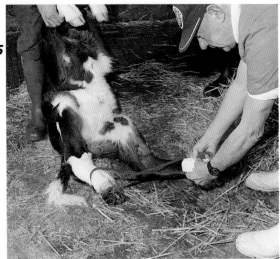

Take the rope, pass one side of the loop over the foot to just above the fetlock, and twist it into a figure of eight (*photo 4*).

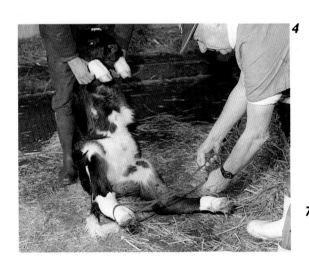

Put the other side of the 'eight' over the other foot, and again up to just above the fetlock (*photo 5*).

Fix the calf's hind legs by placing one foot on the centre of the figure of eight. This will save many a painful kick on the shins (*photo 6*).

After cleaning and disinfecting the area with surgical spirit, insert half a cc of local anaesthetic underneath the skin at the base of the extra teat and leave for at least a minute (*photo 7*).

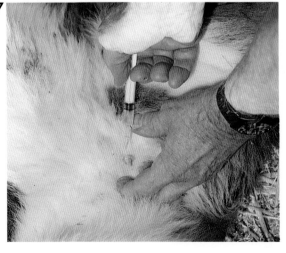

Grasp and fix the teat firmly in the jaws of the artery forceps (*photo 8*). With the forceps, pull the teat outwards as far as possible, and then take a generous 'bite' with the curved scissors around the teat base. Don't be afraid to take a good portion of the surrounding skin (*photo 9*).

This leaves a comparatively large elliptical wound, but it ensures the complete removal of the rudimentary milk sinus which, if left, would produce subsequent leaking (*photo 10*).

Finally, dress the wound thoroughly with antibiotic. Any infection in this area, even in a young calf, can ruin the potential milk production (*photo 11*).

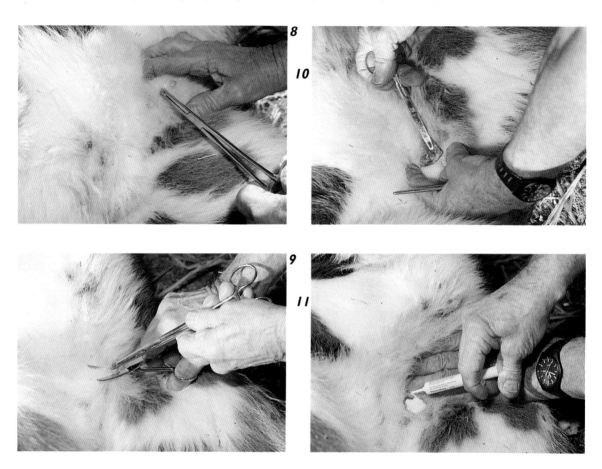

8

10

9

11

32
Calf Castration

Ten to 12 weeks is the best age for castration. By that time the testicles are developed sufficiently to be easily handled *(photo 1)* and the calf is old enough to avoid shock. If the calf is over two months old, a local anaesthetic must be used.

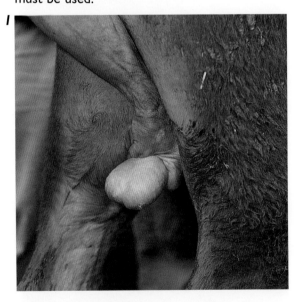

antiseptic and cotton wool and a selection of antiseptic powders and sprays to dress the wounds. Use plenty of non-irritant antiseptic and wash your hands thoroughly with soap and hot water. Remove your watch *(photo 3)*.

DRAWING METHOD

The patient is tied by a halter. An assistant holds the calf's tail. Hot water, soap, a towel, antiseptic and a tray of sterile and antiseptic tools are required *(photo 2)*: two scalpels (one spare in case the blade becomes detached), artery forceps in the unlikely but possible event of haemorrhage, scissors for use with the ligature method, catgut for the same purpose, syringe and needle, local anaesthetic, skin

121

Wash the testicles thoroughly *(photo 4)*. Grip the top of the testicles and turn them towards you *(photo 5)*. The correct site of your incisions so as to achieve perfect drainage is the **base** of the testicles. Soak a cotton wool swab in the special skin antiseptic and prepare the surgical sites *(photo 6)*.

Fill the syringe with 5cc of local anaesthetic. Using a fine sharp needle (supplied by your veterinary surgeon), inject half the local anaesthetic under the skin and the remainder straight into the first testicle *(photo 7)*. Repeat with the second testicle and wait four or five minutes for the anaesthetic to take full effect *(photo 8)*.

4
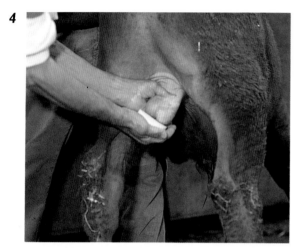

7
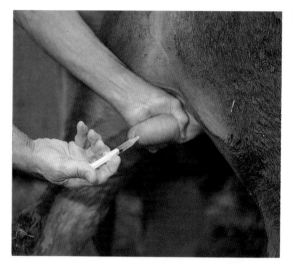

5
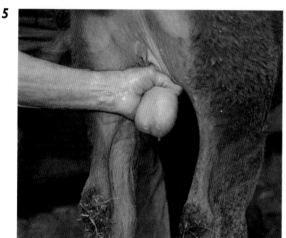

8
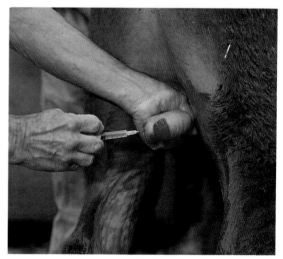

6
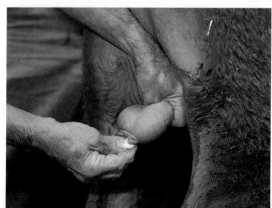

Make an incision transversely across the base of the first testicle *(photos 9 & 10)*. Expose the testicle *(photo 11)*. Withdraw the testicle gently *(photo 12)*. The patient should feel no subsequent pain.

Using the sharp blade of the scalpel, cut

through the membrane that anchors the testicle, taking great care not to cut a blood vessel *(photo 13)*. Separate the membrane as high as possible from the spermatic cord *(photo 14)*.

9

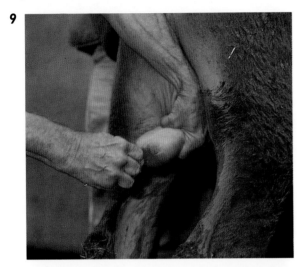

10

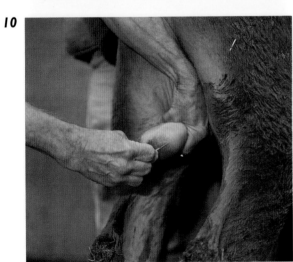

11

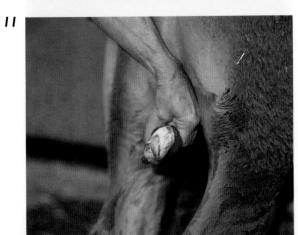

12

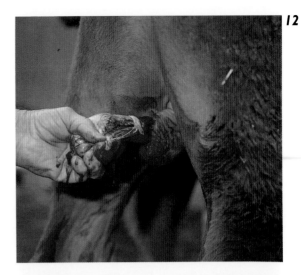

13

14

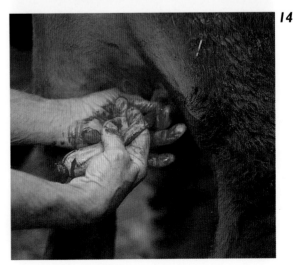

Wrap the cord round the index finger of your right hand (left hand if left-handed) (photo 15) and pull hard (photo 16). The stretching of the cord (including the main blood vessel) produces an elastic contraction when the cord breaks, and this stops the bleeding (photo 17).

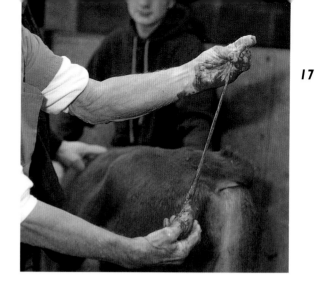

17

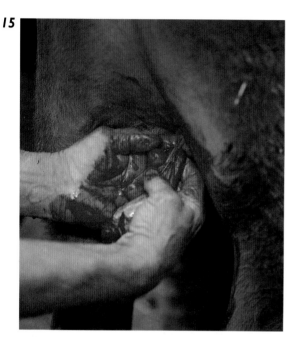

15

Repeat the entire process with the second testicle and dust the wounds copiously with an antiseptic powder. Alternatively you can spray the wounds with an antibiotic aerosol (photo 18). If the castration is done during the fly season it is advisable to use an anti-fly dusting powder (photo 19).

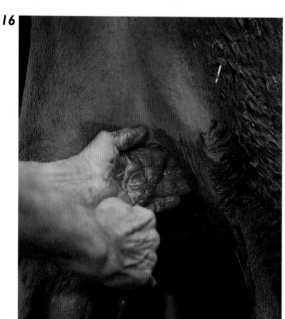

16

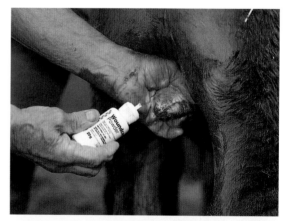

18

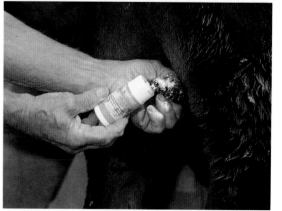

19

LIGATURE METHOD

For older calves an alternative to the drawing method is to use a sterile catgut ligature *(photo 20)*. Tie the ligature tightly in a double knot round the spermatic cord as high as possible, and cut the cord with sterile scissors approximately one inch below the knot *(photo 21)*.

BURDIZZO METHOD

When using the bloodless castration, that is the Burdizzo, prepare the injection sites high up over both spermatic cords *(photo 22)*. The local anaesthetic is infiltrated under the skin and into the cord on each side *(photos 23 & 24)*.

20

22
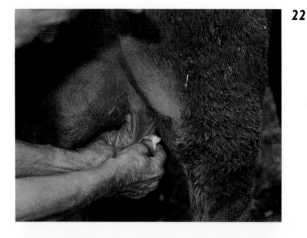

21
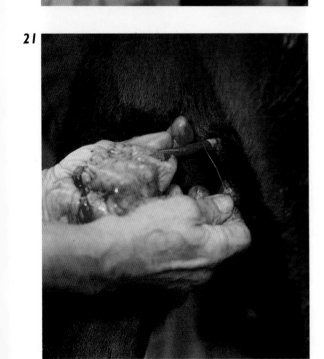

23
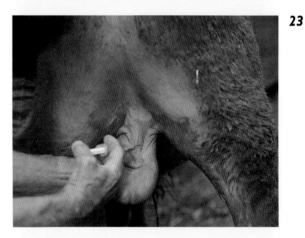

24
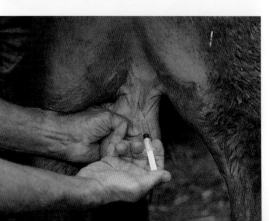

Close the instrument tightly over both spermatic cords *(photo 25)*. Personally I do not like or recommend this method since it produces prolonged pain over several days, whereas in cutting the pain is gone as soon as the operation is completed.

25

Photographs 26-28 illustrate how the bloodless castrator produces its effect. It cuts the cord without cutting the skin. As you can see, it cuts the string without cutting the bank note.

26

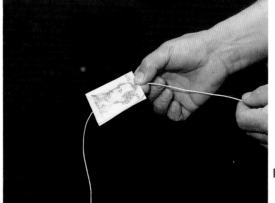

27

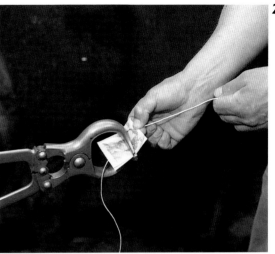

28

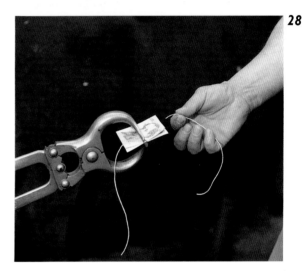

RUBBER RING METHOD

Castration by rubber rings, like that with the Burdizzo, involves a considerable period of pain and for that reason I do not recommend it. It is illegal to use a rubber ring on a calf over one week old.

The rubber rings act by cutting off the blood supply to both the testicles and the sacks in which they are contained. Not only do the rings cause several hours of pain but the ulcerating wounds produced provide a potential field of growth for the tetanus bacillus.

33
Cryptosporidiosis

Recently this condition has caused increasing concern in calf rearing.

Cause
A small protozoan parasite which is related to Coccidia.

Symptoms
The parasite affects the same area as E.coli, namely the lower part of the small intestine: there, when present in numbers, it causes acute dehydration, loss of appetite, scouring (photo I) and death if untreated.

Treatment
Since the Cryptosporidia is related to the Coccidia one would expect the treatment recommended for Coccidiosis to be equally effective, but in the vast majority of cases this is not so and drug therapy, on its own, is often unsuccessful.

In co-operation with your veterinary surgeon, rigidly control the dehydration for as long as necessary and apply the oral sulpha drugs in a completely new and sterile environment.

Prevention
By far the most important factor is to make sure all calves have an adequate supply of colostrum (photo 2).

As a precaution against further cases empty and sterilise the original affected box and leave empty for at least a month.

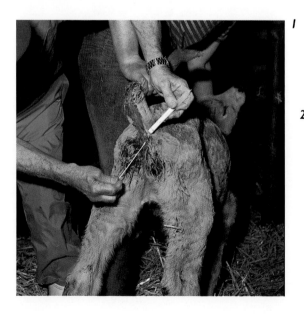

I

2

34
Goitre

Goitre is one of the main causes of stillborn calves. Several years ago I had this problem in my own dairy herd. A veterinary investigation officer from Liverpool University diagnosed and corrected it for me.

Cause
A deficiency of iodine that leads to a shortage of the thyroid hormone thyroxine, which normally stimulates the workings of the body.

Symptoms
The shortage affects the milk yield and the fertility of the cows, and leads to weakly calves that are often stillborn (photo 1).

The attempt to produce the missing iodine increases the size of the thyroid gland to produce the condition of Goitre.

The diagnosis is confirmed by blood examination.

Treatment
Subsidise the dairy ration with iodised minerals.

Prevention
Avoid excess feeding of kale, soya bean, turnips and rape straw, all of which have a low iodine content.

1

Index